Exploring Screen Culture via Apple's Mobile Devices

Exploring Screen Culture via Apple's Mobile Devices

Life Through the Looking Glass

Charles Soukup

LEXINGTON BOOKS
Lanham • Boulder • New York • London

Published by Lexington Books
An imprint of The Rowman & Littlefield Publishing Group, Inc.
4501 Forbes Boulevard, Suite 200, Lanham, Maryland 20706
www.rowman.com

Unit A, Whitacre Mews, 26-34 Stannary Street, London SE11 4AB

British Library Cataloguing in Publication Data
A catalogue record for this book is available from the British Library

Library of Congress Cataloging-in-Publication Data
Includes bibliographic references and index.
ISBN: 978-1-4985-3960-9 (cloth : alk. paper)
ISBN: 978-1-4985-3961-6 (electronic)

♾™ The paper used in this publication meets the minimum requirements of American National Standard for Information Sciences—Permanence of Paper for Printed Library Materials, ANSI/NISO Z39.48-1992.

Printed in the United States of America

Contents

Acknowledgments

First and foremost, thank you to Christina—my wife, sounding board, proofreader, and so much more. No part of this book would be possible without your boundless support and extraordinary insights. Further, I deeply appreciate my wonderful, uniquely interesting children, Rose and Henry. In many ways, this book was inspired by the imagined technological future you will inhabit in your lifetimes.

I want to express my considerable gratitude to the teachers and colleagues who have kindled and guided my curiosity over the years. In particular, I want to thank the communication studies faculty and administration at the University of Northern Colorado for supporting this project.

I would also like to thank the editors at Lexington for their help in the publication process.

I also appreciate the feedback I received from the early drafts of this project at conferences organized by the National Communication Association, Western States Communication Association, Central States Communication Association, and the Media Ecology Association.

Chapter 3 is a revised version of "The Postmodern Ethnographic Flaneur and the Study of Hyper-Mediated Everyday Life" published in the *Journal of Contemporary Ethnography*.

Chapter 1

Surveying the Looking Glass Life

In a Wonderland they lie,
Dreaming as the days go by,
Dreaming as the summers die:
Ever drifting down the stream
Lingering in the golden gleam
Life, what is it but a dream?

—Lewis Carroll, *Through the Looking Glass*[1]

When Alice journeys through the looking glass, she enters a peculiar world in which she is no longer tethered to reality, at least no reality she has previously known. She is in a perpetual state of uncertainty, never quite able to comprehend the unusual new world she navigates. It is no coincidence that Alice's adventures, now 150 years old, are receiving renewed interest with blockbuster movies, mobile applications, museum exhibits (Hartnett, 2015), and even emoji versions (Cronin, 2015). More specifically, the words *looking glass* evoke diverse connotations strikingly relevant to contemporary social life. In a world filled with digital (glass) screens in our pockets, on our walls, on our desks, in our cars, and virtually everywhere else we turn, Carroll's portal, the looking glass, invites parallels today.

Simply typing "looking glass" into Google produces several persistent results. The most literal, and common, meaning of looking glass is merely a mirror or reflective surface. As Carroll so creatively dramatizes, in a mirror, the reflection is reversed, and everything is turned around when compared to "real" or discernable perceptions. In addition to a reflective image, a looking glass is also a frame, isolating a specific portion of reality. In this regard, a looking glass evokes images of nineteenth-century spyglasses and magnifying glasses that transform reality by enlarging the framed images.

Finally, as a common search result, the 1970s rock group named "Looking Glass" had one enormously popular song, "Brandy" (still a favorite on soft rock radio stations and in karaoke bars across America), that told the story of a lonely woman who was a kind of siren call to restless sailors. Like the looking glass, the screens individuals carry in their hands, pockets, purses, briefcases, and backpacks, flip and distort reality beckoning one's attention at every moment.

THE FOCUS OF THIS BOOK

In general, this book explores mobile digital screens as a cultural experience.

Why do people attend to their mobile devices so frequently? What cultural patterns emerge when significant aspects (both in quality and in quantity) of our everyday lives are experienced through the prism of mobile digital screens? In a phrase, this book is my search for the worldview or *media logic* of mobile digital screens. I believe the media logic of mobile digital screens emphasizes the filtering of relevant information for individuals overwhelmed by a culture saturated by screened data. Mobile devices provide a lifeline that sifts through, limits, and simplifies the complexities of rapid, vast, circulating information in postmodern culture. Particularly, simple, game-like applications (which are the most popular mobile applications), with clear rules and numerical outcomes, exceptionally focus, frame, and filter an overwhelming media-saturated culture. Mobile devices such as the iPhone and iPad clarify and constrain the chaos whirling about the hypermediated postmodern experience.

Attempting to write about the essence or heart of a complex, dynamic cultural experience is filled with peril. There are diverse, ever-changing experiences and readers will no doubt imagine and recall exceptions and alternatives to any of my observations or conclusions concerning mobile devices. Nonetheless, I believe I have discovered significant dimensions of the media logic of mobile digital screens. Of course, people can use mobile devices for countless possible (albeit uncommon) functions such as reading the great novels of Dostoevsky, writing thoughtful meditations on the nature of existence, engaging in empathetic dialogues across cultural differences, or even as a paper weight. I am admittedly offering a partial understanding of mobile devices but one that I believe provides unique and under-recognized insights into the everyday experiences associated with mobile technologies.

Considering Neil Postman's (1986) groundbreaking analysis, television involves a pervasive media logic (or in Postman's language, "bias"), namely entertainment and amusement, with distinct expectations. Despite the biases of television, there are still occasional broadcasts of deliberative political debate, comprehensive examinations of global economic policy, and poetry recitations on TV (although very few people watch)—but these broadcasts feel awkward and ill-suited to the medium. Similarly, the way that game shows and sitcoms so perfectly suit the media logic of television, I believe popular applications that frame and simplify the complexities of everyday life via clear numerical outcomes also better suit the media logic of mobile devices. Stated simply, playing Angry Birds, monitoring fantasy football scores, and following trending topics on Twitter just feel ideally suited to a mobile device.

In many respects, this book charts the terrain of hypermediated screen culture (via the case study of the Apple iOS and the iPhone/iPad) and provides strategies for navigating the complexities of a world dominated by digital screens. The overarching goal of this book is to more complexly understand the role of mobile technology in everyday life (via the extended case study of Apple's mobile operating system for the iPhone, iPad, and iPod) and provide readers more nuanced tools to make sense of our collective journey down the rabbit hole. I believe we are at an essential moment in understanding humanity's relationship to technology. Beginning with Apple's App Store's launch in 2008, mobile smartphone applications are approaching ten years of existence. This represents a unique opportunity to inventory the emergent patterns and imagine new directions for interacting with these devices. Understanding the initial practices associated with mobile applications is tremendously valuable, just as understanding the early experiences with television in the 1950s or the initial uses of personal computers in the 1980s offer insights today. Further, the emergent patterns of mobile applications are still in flux, offering opportunities to change course, as described in the conclusion of this book.

CULTURAL ANXIETIES AND DIGITAL SCREENS

Much like Alice's fantastical experiences, today, individuals are a bit feisty, questioning everything they encounter through the looking glass of their digital screens. The signs of an overwhelming ambivalence, confusion, and dread toward ubiquitous technology are everywhere. Popular media, such as talk shows, morning news programs, and situation comedies highlight stories of individuals who admirably "unplug" or "cleanse" themselves from technology (e.g., Cowgill, 2015). The National Day of Unplugging (March 6, 2015) originated by a creative project called "The Sabbath Manifesto" was a human-interest story covered by such major news outlets as ABC News, NBC News, CNN, PBS, and the *New York Times*. Hollywood movies feature killer robots, uncontrollable artificial intelligence, and surveillance states oppressing humanity. Films such as *The Hunger*

Games franchise, *The Transformers* franchise, and *The Terminator* franchises capitalize on movie goers' overwhelming dread and nihilism concerning technology. Specifically, in the 2015 film, *Terminator Genisys*, an all-powerful artificial intelligence system creates a mobile operating system (that looks an awful lot like Apple's iOS) designed to destroy humanity. These indicators strongly suggest that individuals are struggling to make sense of the centrality of technology in their everyday lives.

Like Alice, many Americans seemingly feel a mixture of puzzlement and agitation, lacking tools to make sense of the present moment. People appear to believe that the ordinary details of their everyday lives, particularly time and space, have been jostled about and turned topsy-turvy. I believe individuals are largely grappling with the inescapability of technology—its overwhelming ubiquity feels somehow sinister, monstrous, and uncontrollable. Popular movies and TV shows prominently feature post-apocalyptic worlds (often inhabited by zombies) conspicuously devoid of technology suggesting a desire to imagine a world (even a world infested with zombies) without technology. Apparently, the anxieties of the present moment pivot on the presence or absence of technology.

People intuitively know their lives are radically changing and these devices are at the nexus of these transformations; yet, individuals lack a viable vocabulary to make sense of this new reality. As Lakoff and Johnson (1980) so elegantly described, metaphors help people make sense of complex experiences such as the nature of argument (e.g., as war) or ideas (e.g., as food). Metaphors serve to highlight some dimensions of an experience and mask or downplay other dimensions of an experience. Most importantly, "we define our reality in terms of metaphors and then proceed to act on the basis of the metaphors. We draw inferences, set goals, make commitments, and execute plans, all on the basis of how we in part structure our experience, consciously and unconsciously, by means of metaphor" (p. 158). In the case of technology, the dominant metaphors for mobile devices strongly influence the patterned uses, definitions of problems, and (ultimately) the proposed solutions to perceived problems. In both popular discourse (e.g., newspaper and television reports, entertainment media) as well as everyday talk with family, friends, and students (who, incidentally, *always* have strong opinions about my work concerning mobile technology), two interrelated metaphors dominate the discourses associated with mobile devices: technology as addiction and technology as a waste of time.

Drawing upon a prevailing discourse of the contemporary culture, in order to explain their relationship to mobile devices, people frequently talk about their "addiction" to technology in a taken-for-granted way, never questioning the veracity or usefulness of this metaphor. Internet or technology addiction is referenced in thousands of news stories, popular memes, and even academic publications (e.g., see Schwartz, 2015). The stories are remarkably similar typically describing an *individual's dependency* on digital screens, cataloguing the compulsive practices associated with text messaging, e-mail, social media, video gaming, Web surfing, and the like. The implication is powerful—not only is technology use habitual, but it is also an inherently unhealthy and dangerous choice (or perhaps a genetic or environmental predisposition) for an individual, similar to smoking cigarettes or using drugs. Unfortunately, addiction metaphors fundamentally misrepresent the complex and intersecting constructive and destructive processes (as well as the institutional integration) associated with mobile technology. In many ways, technology as addiction significantly underestimates the broader social implications of mobile devices by emphasizing only individual responsibility/pathology. For instance, most workplaces do not force employees to snort cocaine or compulsively gamble and most public schools do not strongly encourage students to habitually smoke cigarettes or chew tobacco. But, of course, most workplaces and educational institutions now *require* individuals to use digital technology, often on a daily or even hourly basis. In other words, addictive or antisocial activities are not typically integrated into and sanctioned by a culture's dominant institutions (e.g., family, government, corporations, and education). This is not to suggest that mobile technologies are not tied to a diminished quality of life, as I explore in depth throughout this book. But, unlike addiction, the solution (presumably with the help of mobile technology rehabilitation facilities or mobile technology addiction counseling) is not simply for individuals to "quit" mobile technology, a complete impossibility for the vast majority of people (as I explore in Chapter 2). Most importantly, as a figurative addiction, an individual's lack of self-control and moral discipline (and not cultural or institutional forces) are primarily responsible for any destructive consequences (e.g., loss of privacy, diminished freedom, and pervasive meaninglessness) related to digital technology.

Similarly, mobile devices, again in popular discourse and everyday talk, are often metaphorically described as "a waste of time" (see Connor, 2015). Again, the stories are strikingly consistent describing individual's *lack of productivity* due to personal texting, web surfing, social media use, etc. This metaphor is an extension of what Lakoff and Johnson refer to as "time as a resource"—a metaphor built on a conceptualization of labor as quantifiable. In this case, the individual's everyday life is defined in terms of measurable outcomes tied to labor. Stated more specifically, the individual should always be "productive" (producing quantifiable outcomes) rather than "wasting time" playing Angry Birds, checking football statistics, or following a celebrity on Twitter.

Ironically, information technology is a main factor cited in explanations for *increases* in American workers' productivity over the past forty years.

American workers' productivity "grew by about 74 percent between 1973 and 2013" but workers' wages grew at a much slower rate (White, 2015). By staring at a mobile screen all day (even when playing games or checking Twitter), an individual is always on the clock, always mentally and physically tethered to the workplace via e-mail, text messaging's auditory prompts, calendar reminders, and the like. Anytime away from work is defined as "wasted," evoking shame in the worker (and perhaps rationalizing stagnant wages). As Lakoff and Johnson describe, given the underlying logic of this metaphor, even an individual's leisure time (e.g., time with family or travel) is to be defined in terms of productivity and to be "managed wisely." Not only is the metaphor "mobile technology as a waste of time" largely erroneous (in fact, technology greatly enhances productivity and efficiency), but it compels workers to view the most meaningful moments of life—such as time spent with loved ones, experiences with art and nature, community service, and religious rituals in terms of quantifiable outcomes parsed out with carefully scheduled units (rather than fully experiencing these essential moments of one's existence). In Chapter 6, I describe in greater detail the dangers of a culture dominated by the quantifiable efficiency of technique—which, among other things, explains why people are constantly staring at their phones while at family events, immersed in the beauty of nature, and even while sitting in churches, synagogues, and mosques.

The discourses surrounding mobile devices indicate that people are fumbling around confused, a bit like Alice, lacking explanatory tools to make sense of a bewildering new world. Like all metaphors, technology as addiction and technology as wasting time have elements of truth that ring true for many individuals (as all metaphors hide and reveal aspects of reality)—people genuinely feel dependent on technology and feel their time with technology is meaningless. But, I believe these two metaphors provide severely limited explanations for the role of mobile technology in everyday life. Rather than focusing exclusively on the individual's "choices" associated with mobile technology, the *responsibility* should shift toward imbedded institutionalized patterns of everyday life, patterns that require complex reflection and reconsideration.

In this book, I provide readers a wider frame of reference to understand mobile technology at both individual and broader cultural levels to escape the limits of the simplistic metaphors that presently surround these devices. In particular, in this book, I believe I offer a more fruitful set of concepts to explain the functions, processes, and implications of digital devices within the complexities of everyday life. For instance, the dependency that individuals feel toward their mobile devices can be better understood via their cognitive mapping functions (described in Chapter 4) that clarify and focus the individual's attention in an overwhelmingly media-saturated culture (as described in Chapter 2). Further, ludic (playful) flow (described in Chapter 5) more complexly explains the (seemingly meaningless) fleeting pleasures of these devices that evoke ambivalence and shame in individuals. Finally, the literary aesthetics of equipment for living (described in Chapter 6) offers more viable and sustainable solutions for ethically integrating mobile devices into everyday life.

Curiously, while people I encounter are quick to list all the cultural failings associated with mobile technology, these same individuals use such technologies countless times everyday in their lives. Further, in many of their encounters with mobile technology, they *love* these experiences, feeling the pleasures associated with their mobile applications. I strongly empathize with my friends, family, and colleagues. I feel the same bittersweet emotions about my mobile devices—a strange mixture of pleasure, affection, guilt, and shame. This remarkable ambivalence, these wild swings in individual's feelings about this technology, was a central motivation for writing this book. In other words, why do people habitually use technologies that they are so quick to mercilessly criticize and describe with disgust? Further, how can objects that are at the center of an individual's social and professional lives, objects associated with enormous intimacy, power, and pleasure, be a source of such confusion and uncertainty? As devices are complexly intertwined with fundamental dimensions of everyday life, individuals experience the dialectical tug of the joys and sorrows when interacting with these complicated machines. Neither of these polarized feelings is entirely wrong or right—but both should be tempered with an eye toward the complexities of the role of these devices in everyday life. In this regard, in this book, I deliberately avoid the dystopian/utopian binary that has plagued the recent understandings of digital technology. While it seems utterly obvious that technology is neither a panacea for the ills of society with no discernible shortcomings *nor* a wholly destructive force creating only suffering with no identifiable beneficial applications for people's lives, these images (in advertising, movies, and everyday talk) still pervade the discourses of technology.

This book explores the role of mobile technologies in everyday life via the extended case study of Apple's mobile operating system (iOS) for the iPhone, iPad, and iPod. Via a detailed analysis (including numerous extended examples) of the experiences associated with Apple's iOS devices, this book offers an understanding of contemporary screen culture—a culture filled with digital screens. More generally, the book examines how individuals navigate contemporary culture via mobile technologies. Rather than merely outlining the problems associated with a world dominated by digital screens, this book offers a means for understanding screen culture as well as viable solutions to the challenges facing contemporary social life.

As unifying themes, the concepts of *screen culture*, *cognitive maps*, *flow*, and *equipment for living* run throughout the book. I elaborate in more detail on these concepts in separate chapters and case study examples throughout the book. First, the concept of *screen culture* considers the implications of a contemporary American culture that is dominated, both in everyday practice and in institutional processes (e.g., family relationships, work life, and educational methods), by digital screens. Second, as a *cognitive map* that limits and filters experience, the iOS answers the questions puzzling individuals navigating a screen culture: Who am I? Where am I? What do all these images mean (if anything)? Third, I argue that the unique pleasures associated with the iOS emerge from *flow-like* experiences deriving from three characteristics that are integral to the experience of Apple's devices: (a) a clear set of rules, (b) quantifiable outcomes and feedback, and (c) the feeling of control. Finally, the book explores unique mobile applications that, as an artistic experience, provide *equipment for living* by emphasizing spontaneity, self-reflexivity, moral judgments, mystery, creativity, and above all, an expansion of human freedom. Via these four unifying themes, the book traverses the complex experiences of mobile devices in professional, family, entertainment, political (and many other) dimensions of everyday life. Ultimately, the book provides strategies for interacting with these devices, expanding opportunities for more informed and conscious everyday choices when interacting with mobile machines.

DEFINING SOME KEY TERMS

As clarification, I want to define the relationship between three integral concepts addressed throughout this book: screen, operating system, and application. When an individual stares a smartphone or tablet computer, virtually the entire experience (except perhaps a few buttons on the side) is on *the screen*. In this regard, the screen is the broadest term to describe the experience—so I begin here in Chapter 2 in order to provide a context for the present culture. Narrowing our focus further, all mobile digital devices have an *operating system*. As I am interested in mobile devices as an everyday cultural experience, I am using the term operating system in its most general and cultural/conversational sense (rather than in a relatively narrow technical sense of the term). For instance, Apple describes iOS as the "foundation" of the "mobile experience" (iOS, 2016). As Apple states:

> From the moment you pick up iPhone, iPad, or iPod touch, you know how to use it. That's because iOS was designed to be easy to understand. The simple-yet-beautiful Home screen invites you to discover each of the built-in apps and features. So right away, you're browsing your favorite websites, using search to find whatever you need, or effortlessly switching between apps. And no matter

> what you're doing or what app you're using, everything feels easy, intuitive, and even fun. (iOS, 2016)

In this regard, by operating system I mean the overarching experience (described in greater detail in Chapter 4) of a particular mobile device. Finally, an *application* or app is a specific software that typically involves relatively narrow functions and content. Based on the conversational uses of these terms, it may appear at first glance that a screen, an operating system, and an application (and its "content") are easily differentiated, but as an everyday experience, it is virtually impossible to separate them. These three levels are interdependent and intersecting in complex ways. Mobile devices cannot run applications without an operating system. In fact, Apple defines the iOS as the "DNA" of all apps. Further, the screen is virtually useless (except, I suppose as a mirror) without the "interface" of an operating system of some sort. Applications allow users to customize the aspects of their devices, but only within the (in case of Apple, strict parameters) parameters of the operating system. Applications rely heavily on myriad dimensions of the operating system such as location identification, camera, Wi-Fi and cellular data, bluetooth, volume control, and security settings. Much of the communicative practices of a mobile device, such as digital photographs, are experienced simultaneously within the operating system and specific apps. Similarly, are the finger-swiping gestures of the iOS part of the operating system or the application, when I am interfacing with content within a specific mobile app? On a mobile device, data are moving across and between the screen, operating system, and applications constantly. As an everyday experience, individuals rarely notice distinctions between these levels as Apple and other designers of mobile devices deliberately seamlessly integrate the operating system, applications, and content. Recognizing that these distinctions are somewhat arbitrary and artificial, my choice of terminology is simply intended to clarify my level of analysis. I use the term "screen" (such as in Chapters 2 and 3) to reference the qualities associated with all digital screens filling contemporary culture. I reference the operating system (e.g., description of iOS in Chapters 4 and 5) when addressing qualities and experiences associated specifically with Apple's mobile devices. Finally, I reference specific applications (e.g., Twitter, Google Maps, or the Weather Channel Apps) when describing the unique content in a particular application.

OVERVIEW OF THIS BOOK: A ROADMAP

Each chapter of *this book* examines a distinct dimension of mobile technology's place in contemporary cultural experience. In Chapter 2, based upon the mediatization perspective, I examine digital screens within the context of cultural everyday life. Today, daily life within American institutions is immersed in digital screens. Within contemporary mediatized culture, individuals rely upon the automation of computer software such as apps for mobile devices to filter sensory information. In this regard, screens are *relevance processing machines*.

Given the ubiquity of screens in our daily lives, digital technologies present challenges to social researchers or anyone wanting to engage in a more reflective participation in communication. Chapter 3 explores alternative methods of social research that account for the saturation of digital technology in contemporary cultural life. In order to understand how mediatized experiences fit into everyday cultural practices, I argue that people hoping to regain control over digital technology should embrace the mobilized gaze of the flaneur. In the conclusion of the chapter, I offer a number of recommendations that might provide guidance to individuals entering hypermediated cultural contexts.

In Chapter 4, as a case study, I more thoroughly examine the Apple iOS, the operating system for the iPhone, iPad, and iPod Touch, as a variation of Fredric Jameson's cognitive map. Within an overwhelmingly image-accelerated postmodern culture with hyperreal simulation and boundless signifiers, the Apple iOS (re)locates individuals as a cognitive map that filters experience. In this regard, the iOS directs an individual's consciousness by offering clarity and simplicity to postmodern everyday life. Like the edges bordering a canvas or mirror, the Apple mobile operating system literally frames an individual's experiences.

In Chapter 5, I detail the pleasures associated with the iOS, which compels the practices of ludic (playful) flow. The tremendously popular devices utilizing Apple's mobile operating system are ritualistically integrated into everyday life. More specifically, many mobile applications (e.g., social media, gaming, and fitness apps) "gamify" everyday time and space via clear rules and numerical outcomes, even producing a performative, trance-like state.

In the final chapter, I reconsider the problem of technology (i.e., the logic or worldview of *technique* inherently limits human freedom) originally articulated by Ellul, within the media-saturated environments of contemporary culture. More specifically, the interface with the iOS devices encourages highly rationalistic and standardized cultural practices focused on efficient functionality and quantifiable outcomes. As an alternative, or at least a rupture or disruption in technique, I explore the potential of aestheticized applications for Apple iOS devices as a variation of Burke's concept of equipment for living. The chapter explores unique mobile applications that (unlike the vast majority of mobile applications) transcend technique via aesthetic experiences emphasizing self-reflexive moral judgments.

HOW TO READ THIS BOOK

This book is intended to provide representative observations based on the method described in Chapter 3. In many respects, this is my personal journey (as a middle-class American man) with mobile technology. My exploration of mobile technology began over a decade ago in 2004 (well before the iOS existed) via detailed studies of the iPod and (the now-defunct PDA or "personal digital assistant") Palm Pilot. Over these years, I have thought extensively about mobile technology, especially devices' relationship to my everyday life. The examinations of my experiences are intended to both draw general conclusions about mobile devices and also model a way to self-reflexively understand technologies that fully immerse everyday life. While the specifics of these devices have an expiration date (as the Palm Pilot has become antiquated with the smart phones' emergence), this book's method/process of approaching technologies intimately integrated into the time and space of daily life can remain useful well into the future. I encourage readers to draw upon the observations in this book to become more mindful of the role of digital technology in their lives, within their own unique cultural experiences.

I strongly believe self-reflexivity is the cure to the anxiety and ambivalence that currently surrounds mobile technology. In this book, self-reflexivity, a term used in varied fields such as literary theory as well as ethnographic research, refers to a kind of self-awareness or "consciousness turning back on itself" (Siska, 1979, p. 285), distancing oneself from the mindless habits of everyday life. I like to imagine self-reflexivity as a kind of out-of-body experience (a bit like a ghost in an old horror movie) looking at oneself while floating from above, thinking about the movement and behavior of everyday life. Much of what individuals do with technology is in fact unconscious routine behavior perhaps best represented in the Pavlovian reach for the phone when hearing the iconic text message "ding" (sans the salivation, in most cases). The primary goal of this book is to help evoke greater self-reflexivity by describing the practices and functions associated with mobile devices. By embracing self-reflexivity, individuals can mindfully consider the purposes and behaviors associated with the mobile applications they choose to engage on their devices.

As a starting point, I want to raise two questions that have perplexed policy makers, technology analysts, and designers of public service campaigns. First, despite the widespread knowledge of the dangers of texting (and using one's phone in general) and driving, why do people continue to use their mobile phones while driving automobiles? In fact, while 98 percent of drivers believe the practice of texting and driving is dangerous, "nearly 75% say they do it anyway" (Worland, 2014). But, the issue is not simply driving, as public health researchers also warn against the dangers of smart phone use while walking and performing other everyday tasks (Carroll, 2015). The consequences of these habits range from the comic (such a countless YouTube videos of people walking into walls and trees while using their mobile phones) to the tragic (such as the lives lost or permanently damaged by automobile accidents). Thousands of deaths per year are associated with distracted driving due to smart phone use and, despite concerted public service campaigns designed to prevent distracted driving, there has been no decline in using mobile devices while operating vehicles (US Department of Transportation, 2016). A number of theories explaining the propensity to text while driving have emerged, such as the aforementioned "addiction" associated with the release of dopamine when receiving text messages. Less understood are the cultural conditions, as well as the design structure (in this case the iOS), that evoke this apparent pleasurable anticipation/reward experience when using mobile devices. Most importantly, simply shaming individuals (via public service campaigns) who text and drive has proven ineffective, suggesting that other cultural conditions must be accounted for in order to alter the patterns that induce and sustain this behavior.

Second, after years of warnings from privacy advocates concerning the loss of freedom associated with revealing private personal information on social media, via search engines, and the like, why do Internet users continue to willingly provide personal, private information to Facebook, Google, etc.? Much like texting and driving, Internet users seem quite aware of their loss of privacy while using search engines and visiting social media sites with 91 percent of adults surveyed agreeing or strongly agreeing that consumers have lost control over how personal information is collected and "used by companies" (Madden, 2014). Privacy experts are sometimes perplexed that individuals continue to frequent Facebook and Google despite an apparent awareness that these companies are collecting, storing, and aggregating private information. Much like with the addiction metaphor, discussions of privacy often return to individuals' responsibility for protecting their privacy without accounting for the role of broader institutional and cultural forces.

By understanding the deeper cultural experiences associated with mobile devices, via self-reflexivity, individuals can disrupt routines and patterns that are apparently harmful both personally and socially. In this book, I begin at the broadest cultural level, examining the place mobile devices play within the digital screens of a media-saturated culture. By narrowing the field of vision of an individual, mobile devices filter sensory data providing only relevant information (as I explore in Chapter 2). Once filtered, more specifically, the iOS experience provides essential meanings for the postmodern subject (via cognitive mapping) while evoking flow-like pleasures via gamification. In this regard, symptoms of the media-saturated culture, such as texting while driving and the loss of privacy, can be understood far more complexly in light of my analysis (I return to texting while driving and privacy concerns in Chapter 6). While an individual is certainly partly responsible as a decision maker, in order to truly alter these destructive routine practices, a number of other factors should also be addressed, including the demands of a culture dominated by screens (and the underlying meaninglessness for the individual), institutions demanding attention to screens, software designers imbedding game structure into mobile applications, and a society immersed in a logic of efficiency/productivity. In this regard, this book is designed to change the conversation about mobile devices and everyday life, opening up a more robust understanding of the role these devices play in today's culture. In the process, any number of emergent habits associated with persistent problems, such as texting while driving and protecting private digital information, can be addressed with thoughtful public and personal conversations that can spur enduring solutions.

At the end of both *Through the Looking Glass* and *Alice's Adventures in Wonderland*, Carroll reveals that Alice has been asleep with only fuzzy, fleeting memories of a fantastic adventure. Of course (as Chapter 2 explores), today, we cannot expect to wake up from the dream of ubiquitous digital screens. We must learn to live our lives through looking glass.

NOTE

1. Epigraphs cited from: Carroll, Lewis. *Alice's Adventures in Wonderland* and *Through the Looking-Glass*. New York: Grosset & Dunlap, Inc, 1946.

Chapter 2

Relevance Processing in Twenty-First-Century Screen Culture

A few years ago, sitting in a generic hotel room in San Francisco, I was watching a video (on my laptop computer) with my daughter titled "*Dora the Explorer* and the Lost City." Like all episodes of the enormously popular children's television show, the video opens with a shot of the living room of a family's home and pans toward a personal computer on a desk. The camera locks on a computer screen (at which point it shifts from live action into animation) and suggests that the show becomes an interactive computer program. In this specific episode, Dora and her monkey sidekick, Boots, are searching for a lost teddy bear that they ultimately discover is in the Lost City. While searching for the Lost City, Dora and Boots encounter a magical screen (simply referred to as "screen") who directs them along their way with relevant information. Upon meeting the screen, Dora says, matter-of-factly, "Let's listen to the screen for directions." Throughout the episode, whenever they need information, the interactive screen (similar to a hovering iPad or interactive kiosk at an airport but with eyes and a talking mouth) magically appears and provides important advice for Dora and Boots. Sitting in the placeless environment of a hotel room, looking at a computer laptop depicting an adventurous bilingual girl (who inhabits a digital computer program) and her monkey directed by ephemeral digital screens, I looked at my daughter and thought to myself, "Welcome to our new reality."

In this chapter, I introduce a central question that inspired this book: Despite a pervasive discourse articulating an almost uniform cultural preference for "face-to-face" communication (embodied, co-present experience), why do people consistently attend to their digital screens in lieu of "live" or nonmediated communication in their everyday lives? American culture is dominated, both in everyday practice and in institutional processes, by digital screens. Further, the media logic of the digital screen is defined by a single concept: relevance. The intersection of digital media's interoperability (i.e., seamlessly movement of data across devices) and diversity of scale with the inherent invisibility of screens have fostered a ubiquitous mediatized logic of relevance that is hidden in plain sight. Immersed in digital screens, individuals tune experience by *screening* relevance in everyday life.

WHAT IS A DIGITAL SCREEN?

The concept of the digital screen is composed of two interrelated ideas—the *digital* and the *screen.* In simplest terms, "the digital" refers to numerical representation via binary code (Manovich, 2001) or the proverbial ones and zeros that are associated with all information technology. Evens (2003) explained the unique qualities of the digital cogently:

> Digits provide a universal standard that is easily transportable, exactly duplicable, and broadly applicable. . . . The digital thereby amplifies technology's powers in many ways, stamping the world with an order that is both highly malleable and utterly consistent. (p. 53)

Converting communicative messages into digital code provides astonishing flexibility in terms of scale/size, form, interactivity, and interoperability. For the purposes of this analysis, the mathematical consistency of digital code enables media objects of incredibly diverse size and shape (e.g., high definition or HD televisions, smart phones, personal computers or PCs, sports stadium jumbotron screens) as well as the ability to move content (e.g., video, audio, documents, and webpages) seamlessly between these media objects. The diversity and interoperability of digital media devices are experienced by individuals almost exclusively via a familiar cultural object, "the screen."

Although a ubiquitous term in contemporary culture (e.g., flat screen, touch screen, computer screen, movie screen, and TV screen), the notion of a screen is actually relatively difficult to define. In abstract conceptual terms, screens offer visual representations of something (actually, almost anything). As a significant cultural object, what is peculiar about a screen is that it is intended to *fade into invisibility.* As Introna and Ilharco (2004) explained:

> When trying to describe a screen, a computer screen or a television screen, we immediately note that we never seem to look at a screen, as a "screen." We rather tend to look at screens by watching what appears on them. What seems evident when looking at a screen is the information presented on that screen—the text, images, colours, graphics, and so on—not the screen itself. (p. 224)

When described in abstract terms, the "digital screen" sounds a bit ambiguous and even science fictional—*an obscured visual surface utilizing the principles of binary code*. Despite its relative newness and ostensible conceptual murkiness, the digital screen takes many familiar ever-present forms. Immediately, to any participant in an information society, apparent examples include the mobile phone, the HD television, tablet computer, mp3 player/iPod, and the notebook/desktop computer monitor. Digital screens also take many perhaps less-noticeable forms such as automatic teller machines (ATMs), airport kiosks, digital cameras, global positioning system (GPS) devices, digital billboards, and bus/train advertisements. With the recent transition to digital television and film, analog screens are quickly drifting toward obsolescence (nonprofit second-hand stores such as Goodwill no longer even accept analog televisions as donations). The unique flexibility and interactivity of digital screens has fostered a dizzying explosion in daily practices that now saturate contemporary culture.

Screens Big, Screens Small, Screens Everywhere

Growing evidence suggests that Americans live in a culture dominated by digital screens, as "Americans now own four digital devices on average, and the average U.S. consumer spends 60 hours a week consuming content across devices" (Nielsen, 2014). The *average* teenager now spends more time on digital devices than they spend sleeping (Wallace, 2015). Ninety-two percent of Americans own a cell phone, with smartphone ownership "nearing the saturation point with many groups: 86 percent of those ages 18–29 have a smartphone, as do 83 percent of those ages 30–49 and 87 percent of those living in households earning $75,000 and up annually" (Anderson, 2015). Further, 76 percent of Americans own a computer, with the greatest statistical increases in laptop ownership (Smith, 2010). A group of researchers from Ball State University's Center for Media Design estimated that "adults are exposed to screens—TVs, cellphones, even G.P.S. devices—for about 8.5 hours on any given day" (Stelter, 2009). More importantly, these devices fill the most mundane moments of Americans' lives, as "Fully 93 percent of 18–29 year old smartphone owners in the experience sampling study used their phone at least once to avoid being bored" (Smith, 2015). Even with the enormous numbers of people utilizing personal computers and mobile phones, television remains ubiquitous. Today, an average American home has more televisions than people (Bauder, 2006) and there are now more televisions in the United States than people (Starr, 2009). At restaurants, airports, coffeehouses, college campuses, and even city streets, digital flat screens clutter the walls of public spaces. Put simply, Americans love their screens and watch and/or interact with them more every year (with no end in sight). Consider, for example, Samsung, LG, and Whirlpool all produce refrigerators that include a digital screen imbedded in the door (for wireless Internet, television, apps, and DVDs).

Observing the astonishing proliferation of screens in contemporary culture, Livingstone (2009) provocatively argued that "information and communication technologies now mediate every dimension of society" (pp. 1–2). Deuze (2011) extends this line of reasoning by concluding: "Our life is lived *in*, rather than *with*, media" (p. 138). Screens are so commonplace and seamlessly integrated into everyday experience they become obscured, even invisible, to individuals—a fact not terribly surprising considering the aforementioned inherent obscurity of all screens. In this regard, screens (and their media logic) are simultaneously everywhere all the time and hidden in plain sight. As individuals are immersed in a sea of mediation, "we can no longer step outside of the global communications flow to find a solid fulcrum for critique. There is no more outside. The critique of information is in the information itself" (Lash, 2002, p. 220). In a hypermediated culture, individuals cannot escape digital information. Like Jim Carrey in the film, *The Truman Show* (Deuze, 2011), there are no spaces beyond digital media nor are there opportunities for reflexivity outside the world of screens. In this chapter, I want to introduce a central question of this book—Assuming that the ubiquity and banality of screens marks a significant part of our everyday social and individual realities today, what are the sense-making processes for individuals enveloped in it? In my investigation, I intend to decode and demystify the digital screen within the context of cultural everyday life (Lundby, 2009).

I am using the term *sense-making* based on the phenomenological understandings of everyday-mediated experiences. Strongly influenced by the work of Heidegger, Scannell (2000) problematizes the notion of a uniform mass "audience" advocating a move toward communicative practices associated with issues of identity, personal experience, and temporality. Similarly, Introna and Ilharco (2004) conceptualize the "screen" (televisions, computers, etc.) as an "intentional object of our attention" (p. 227). The common thread among these phenomenological approaches to mediated everyday life is the conscious media-related practices of individuals navigating the complex landscape of contemporary culture. Initially, the diverse and contradictory meanings attributed to digital screens might be teased out by surveying the terrain of digital screen culture.

Digital screen culture involves complex layers of digital mediation—what I deem, hypermediation. For instance, in a kind of infinite loop of mediation, media consumers now watch television and computer screens that depict people who are also staring at television and computer screens (and so on). At times, particularly during election coverage, CNN will portray news anchors such as Wolf Blitzer or Anderson Cooper standing in front of as many as nine screens (and of course with laptops directly in front of them), highlighting "breaking" news events of the day. On ESPN and TNT Sports, analysts stand in front of giant touch screens (which are literally larger than life, as the images of athletes dwarf the broadcasters) interacting with clips of sporting events. As absurd as all these layers of mediation may initially seem, viewers do not seem terribly surprised. The television sets of sportscasters and news anchors offer a microcosm of diverse cultural environments engulfed with digital screens. The message is simple: what is on the screen is what is important.

Digitization has enabled screens to become astonishingly small *and* large. Essentially, digital screens are tailor-made for every conceivable specific function or context (e.g., placed on the back of an airplane seat, mounted to the ceiling of a mini van, and put in an individual's pocket). The tremendous popularity of Apple's iOS devices (iPod, iPhone, and iPad) and other smartphones/tablet computers indicates that consumers desire mobile gadgets of varying sizes to suit unique purposes. Further, the jumbo screen in the Dallas Cowboys' Texas stadium, 180 feet long and 50 feet tall, demonstrates that digital screens are capable of reaching astounding scale. Due to its size and HD clarity, spectators report that it is difficult to *not* look at the screen during live sporting and music events at the new Texas stadium, even when seated near the playing field or stage (LeBreton, 2009). Further, the digital screen in the planned football stadium in Los Angeles will be 50 feet tall and 120 yards long, twice the size of the screen in Dallas (SI Wire, 2016).

In addition to tremendous variation of scale, the diversity of digital screens has introduced a new phenomenon—interacting with multiple screens simultaneously.

The buzzword "multi-tasking" represents the pervasive process of navigating many screens concurrently. For instance, marketing research indicates that people increasingly surf the Web or post to social media while watching television using (in marketing terminology) a "second screen" (Atler, 2014). The simultaneous use of multiple screens is a by-product of the screen's diversity of purposes. Remarkably, screens transcend contexts and functions within everyday life. Screens are now fully integrated into an individual's work, interpersonal, and leisure time. As a National Public Radio (NPR) report states, in a bizarre irony of twenty-first-century life, Americans often "escape" their work screens by retreating into their home screens (Blair, 2009). In fact, an increasing number of Americans spend more of their income on screens (e.g., mobile phone, Internet, cable television, and streaming services) than on groceries (Darlin, 2005). Further, recent research indicates that a growing number of people across demographic groups (not only teenagers) spend more time on their digital devices than they spend sleeping

(Timberg, 2015). These trends indicate that digital screens are viewed as essential to everyday life, like shelter, sleep, and food.

Cable television and Internet provider Comcast recently launched an advertising campaign featuring "Comcast-town"—a computer-generated environment featuring screens in every imaginable public or private space, a town in which screened images from movies, television, the Internet, and video games come to life as one walks or drives the streets of the virtual town. With its upbeat soundtrack and vibrant colors, Comcast depicts a wondrous culture (a wonderland) dominated by screens, in which a city has screens on every open wall or available open space. In this respect, there is an obvious hegemony of the digital screen that will likely be familiar to most readers—screens are unavoidable and resistance is futile.

From the examples highlighted above, individuals, despite a pervasive ambivalence about the centrality of technology in contemporary culture, perceive the ubiquity of screens as both normal and irresistible. Consider, for instance, a Sprint advertisement (representative of the marketing of mobile phones and wireless carriers) that beckoned the viewer to imagine their childhood dreams of space travel and time machines. The television commercial then described a mobile phone as a "magic screen . . . that entertained you whenever you wanted, that held your favorite music, pictures, stories, and helped you find your way home. A screen that brought all of this to you wherever you are at the speed of light. That was your dream." As I explore further below, screens have a powerful pull on an individual's consciousness and attention. Further, the tremendous interoperability across screens of diverse size and shape indicates meaningful *connections* (typically referenced via the buzz word convergence) between these screens, rather than discernible *distinctions*. In other words, while the piecemeal social scientific approach to these issues (e.g., examining mobile phones as distinct from personal computers and studying social networking as distinct from text messaging) provides some insight into everyday-mediated life, as I continue to examine the literature in this field, I feel there is something bigger afoot in American culture with respect to digital screens.

MEDIA LOGIC AND MEDIATIZATION

Within the communication discipline, the notion of mediation or "mediated" communication is foundational to the understanding of a wide range of communication processes associated with technologies/media such as the telephone, television, and (more generally) language. In essence, mediation refers to the "channel" in the sender-receiver or cybernetic model of communication. For instance, in their groundbreaking anthology, Gumpert and

Cathcart (1986) frame the relationship between interpersonal communication and technology via the concept of mediation represented in the channel "interposed" between communicators. In this conceptualization, the focus is relatively narrow: isolating the role of a communication technology primarily in interpersonal message exchange.

I am aligned with scholars who feel the cultural context is deemphasized by a mediation conceptualization of technology (see Carey, 1989). A compelling complementary concept is mediatization. Mediatization shifts the attention from sender-receiver (i.e., one message sent between two people) flow of information (cybernetics) (Craig, 1999) within a specific context toward the long-term institutional, social, and cultural implications associated with the *media logic* of technologies in the everyday lives of cultural participants. As Hjarvard (2008a) explains, "Mediatization is the process of social change that to some extent subsumes other social or cultural fields into the logic of the media" (p. 14). In many ways, this is the focus of the present project: the cultural dimensions (i.e., mediatization) of digital screens.

The cultural dimensions of digital screens are represented in extensive and diverse examples. A version of Benjamin's (1999) phantasmagoria pervades contemporary culture. Phantasmagoria, deriving from the two words "phantom" and "agoria," represents "the illusionistic screen entertainments that manufactured phantoms out of light" as well as the agoria or marketplace (Friedberg, 1993, p. 82). Benjamin often described these "phantoms of light" as invoking a dream-like state (Buck-Morss, 1989). In this regard, phantasmagoria is more than merely advertisements or shop window displays—it envelopes the urbanite in the marketplace of illusory commercial media as "Benjamin believed that objects were commodified into dream images or phantasmagoria that became lodged in the collective consciousness of a culture" (Cvoro, 2008, p. 91). I think it is reasonable to suggest that Benjamin, relying on words like "phantoms" and "illusions," characterized phantasmagoria as conjuring a kind of magic.

Supermarkets are experimenting with digital screens in the store's shelves offering video advertising near products such as a Cheerios' commercial running in the cereal aisle (Elliott, 2010). American toy icon Barbie is now outfitted with a wide range of digital screens such as smart phones and notebook computers. Mattel also has a line of Barbie-related cell phones, handheld computers, interactive games, mobile applications, and video game systems marketed to "tween" girls. Digital screens are remarkably normal and are regularly integrated in some previously unexpected places, such as a text-messaging cartoon toucan (in the jungle, no less) on a bottle of Suave children's shampoo.

In the contemporary world of phantasmagoria, screens depict the dreams of consumerism. Screens are filled with commercial media products—in one form or another, almost everything on screens is for sale. In Lanham's (2006) estimation, in this "economy of attention," screens must compete for our attention: "The screen works differently from the page. Words don't stay put. They dance around. Images play a major role and they move too. Color is everywhere. And sound, too, spoken and synthesized . . . the digital screen depends on an economics of plenty" (p. 20). As Lanham explains, we no longer live in an economy of "stuff" (physical goods) but rather an economy that relies on a rhetorical process ("fluff"): drawing our attention. Our attention is pulled and prodded, provoked and poked as we move about our everyday lives. In other words, in the language of mediatization, attention is central to the media logic of digital screens.

Schulz (2004) explains that, from the perspective of mediatization, "Media use is woven into the fabric of everyday life; the media pervade the professional sphere, the economy, culture, politics, and the public sphere. Media activities and non-media activities amalgamate" (p. 89). In this regard, mediatization should not be confused with traditional media effects or technological determinist approaches due to its unique emphasis on everyday life, especially the micro level of social interaction (Lundby, 2009). Similarly, mediatization avoids the trap of a "single medium" and instead embraces the obvious convergence(s) that complexly integrate technologies with everyday life, especially in the social construction of institutions such as family and education (Krotz, 2009).

As an abstract theoretical concept adopted by a wide range of academic disciplines, mediatization has been conceptualized with a predictably diverse set of approaches and applications. For this book, I am drawn to the postmodern applications of the term, especially the notion of an underlying "media logic" within communication technologies (Couldry, 2008). Hjarvard (2008b) summarized: "the term 'media logic' refers to the institutional and technological modus operandi of the media, including the ways in which media distribute material and symbolic resources and operate with the help of formal and informal rules" (p. 113). Stated simply, a media logic "formats" social experience. I am particularly compelled by mediatization's simultaneous emphases of the macro (broad institutional transformations) and micro levels (details of everyday life) of social interaction. From this perspective, as a technology becomes ubiquitous within a culture, people orient toward their social lives in new ways (Silverstone, 2002). In short, mediatization helps us see how technology becomes intimately intertwined with everyday lived experience.

More specifically, the temporal and spatial dimensions of everyday life are complexly interconnected with digital screens. Time and space are fragmented and displaced as individuals are decreasingly "grounded" or tethered to a kind of physical shared reality. For instance, media spectatorship involves a pervasive "temporal displacement" (Friedberg, 1993, p. 125) as digital texts are viewed out of order, such as skipping around on a DVD/DVR, catching a scene or two of a movie on cable TV, "binge" watching an entire season on Netflix over a weekend, and seeing a clip (out of context) of a television show on YouTube. Further, wireless communication compresses and desequences time, creating new social practices (Castells, Fernandez-Ardevol, Qiu, and Sey, 2007). Caron and Caronia (2007) conclude that mobile devices are used to construct meaningful moments from "no-where places" and "no-when times" (p. 38). When interacting with digital screens, individuals shift moment to moment between "movie/video game/music time" (bodyless, mediated, desequenced) and "real time" (physical, embodied, linear). In order to explain the relationship between space and technology, Sheller and Urry (2006) conceptualized "technospaces" that "enrol people, space, and the elements connecting people and spaces, into socio-technical assemblages" (p. 9). These observations bolster the notion of "multilocalization" (Caron and Caronia, 2007, p. 17) as individuals can be *moving between* spaces (and times) and simultaneously be in multiple "places" (and times) via digital screens. Taken together, these examples suggest that ubiquitous digital screens should not be considered detached or separated from the movements and temporality of everyday life; rather screens should be thought of as thoroughly integrated into everyday time and space.

Tuning Screen Culture: The Media Logic of Relevance

Culture is composed of habitual practices. Daily, cultural participants, sometimes unwittingly, follow consistent patterns based upon identifiable norms. Technology is one cultural formation that involves "a whole complex of practices, assumptions, habits, competencies, and so on" (Krotz, 2009, p. 27). For Coyne (2010), everyday life is defined by a meaningful repetitiveness as "Repetition is never 'mere repetition.' Repetition provides social order, coordination, synchronization, and tuning in various forms" (p. 84).

In a screen culture, the individual's attention is pulled and prodded, provoked and poked while moving about everyday life. The ubiquity of digital screens has prompted the publishing of books such as *Distracted: The Erosion of Attention and the Coming Dark Age*, *Hamlet's BlackBerry: A Practical Philosophy for Building a Good Life in the Digital Age*, and *Is Google Making Us Stupid?* The general dystopic tenor of these popular accounts is amazingly similar: digital screens are overwhelming the senses, reducing attention spans, and basically destroying human brains, human relationships, and the society at large. Almost daily reports in the popular press about the declining quality of life due to omnipresent technology further suggest a growing "common sense" concerning the destructiveness of ever-present screens within everyday life (e.g., see NPR Staff, 2010). This conventional wisdom (introduced in Chapter 1) concerning the proliferation of screens begs a key question: If these screens are so terrible for an individual's quality of life, why do so many people attend to these screens so frequently?

Coyne (2010) offers a compelling alternative to this pervasive common sense that paints the individual as a fool who irrationally (and, as discussed in Chapter 1, "addictively") stares at digital screens. Coyne's work draws attention toward the ordinariness of ubiquitous digital technologies, not falling into the trap of characterizing these technologies as revolutionary, extraordinary, and/or dystopic. For the people interacting with screens everyday they are remarkably *routine* and largely seamlessly (and obscured/invisibly) integrated into everyday life. On the other hand, when the "seams" are exposed (p. 19), such as when a machine malfunctions or works slowly, it reveals much about the mediatized routine practices of everyday, life. Following Coyne's logic, the reason people adopt digital devices at such an astonishing rate is a function of temporal/spatial calibration, social coordination, and adjustment. This turns the common sense (i.e., individuals are self-destructively and foolishly "addicted" to their devices) swirling within American culture on its head—*without* these devices, the time and space of everyday life become disharmonious and in a state of discord—as anyone who has lost their mobile phone realizes quite quickly. As Jerry Seinfeld recently joked, "There is no safety, there is no comfort, there is no security for you in this life anymore unless when you're walking down the street you can feel a hard rectangle in your pants" (Jerry Seinfeld, 2014). In other words, rather than simply destructive or alienating, screens are increasingly *relevant* to the average individual.

In concert with Coyne's observations, the compelling conclusions offered by Introna and Ilharco (2004) sparked a chain of analysis for my work: "the essential way in which the screen is what it is in-the-world is addressing or mediating our being in the world by displaying or presenting relevance" (p. 230). Drawing upon this conclusion, the term "screen" is best understood as both a noun and a verb as digital screens *screen* the individual's perceptions of the world. Although the idea that public discourse is performed via screens (e.g., television, newspapers, and Internet) has received some attention (DeLuca and Peeples, 2002), the screening of individuals' everyday lives via digital devices has remained relatively unexplored.

For individuals within the contemporary mediatized culture, screens are *relevance processing machines*. In screen culture, people are expected to make split-second decisions about vast amounts of screened and nonscreened sensory data. Individuals must, in a word, make countless decisions about what perceptual information is *relevant*. In this complex sensory environment, relevance becomes the overriding value of digital screen culture. Within the media logic of mobile digital devices, sensory information must be immediately relevant to apparent goals (the ideas of efficiency and technique are explored in greater depth in subsequent chapters). The instantaneous decisions individuals make concerning relevance are based on routine habit aided by the filtering of interoperable digital devices of varying size and shape. The speed and efficiency of information flow within digital screen culture demands relevance filtering mechanisms, namely software (e.g., "apps" for smartphones and tablet computers). People, in Coyne's terms, simply become "tuned" to their screen-saturated environment, relying upon the automation of computer programs to filter relevance. Further, with only split seconds to shift attention, people rely heavily on sensation to determine their choices—simple, pleasing, familiar images and sounds draw an individual's attention.

Mediatization processes encourage specific orientations to relevance. Relevance within mediatized screen culture appears to be defined, based upon an overriding media logic, via a specific set of filtering criteria. The criteria for evaluating relevance intersect with the closely related values of *efficiency* and *fleeting pleasure*—what is relevant is what is *immediately* useful and sensually stimulating/pleasing. For instance, in the iTunes App Store, the most popular apps consistently fall into a few categories: social networking (e.g., Facebook and Twitter), entertainment media (e.g., Pandora and Netflix), information "screening" (Webmd and Yelp), and games (e.g., Angry Birds). Apps filter data about family/peers, entertainment media, medical information, and so on for relative relevance. As I explore further throughout this book, the everyday practices associated with digital media continually suggest that fleeting pleasure and immediate/efficient function are among the preferred criteria for judging relevance.

From a mediatization perspective, relevance is what appears significant or valuable *within the logic of digital screens.* A media logic, as I explore in Chapter 6 in terms of Ellul's concept of technique, has an underlying set of values that foregrounds "relevant" details of reality and blurs or masks "irrelevant" details of reality. In the process, the media logic of digital screens can potentially conceal meaningful and ethical dimensions of everyday life that are (literally and figuratively) "screened" from an individual's consciousness. When fully immersed in a culture filled with digital screens, I realized that the values of efficiency and fleeting pleasure are foregrounded in my consciousness while the values of community, empathy, and reciprocal human relationships are largely obscured. In Chapter 6, I consider possible practices for revealing the hidden dimensions of everyday life. In the following section, I explore further an individual's consciousness when immersed in the logic of digital screens. Further, this initial discussion is intended simply to introduce readers to the shear immersiveness of the logic of digital screens—the specific details of this media logic (particularly the clarity, efficiency, and fleeting pleasure associated with Apple's mobile devices) are examined in greater depth via the discussions of cognitive maps, flow, and technique later in this book.

Screening Relevance in Everyday Life

For an individual immersed in the contemporary digital media logic, if the screen offers relevance, the "nonscreen" has come to offer (relatively speaking) irrelevance. The nonscreen might be best described as "liveness" (Auslander, 1999) marked by physically copresence performances (i.e., immediacy). Nonetheless, while "live" and "screened" may appear to function as a strict dichotomy, in practice, contemporary everyday life is largely experienced on a continuum of relative levels of mediation. For instance, Auslander observes that "Live performance now often incorporates mediatization to the degree that the live event itself is a product of media technologies" (p. 25). Auslander insightfully recognizes that the line between "screened" entertainment and "live" entertainment has greatly blurred with Broadway plays like *Beauty and the Beast* and live reproductions of music videos in rock concerts which are both "filtered" through their "own mediatized reproductions" (p. 35). The individual navigating a mediatized culture must make sense of a world filled simultaneously with both screened and nonscreened information (information often woven together), entertainment, sociality, and so on. Consider now commonplace moments such as when friends huddle together to view photographs on a mobile phone or when families stage "photo-ops" designed to post on Facebook—in these moments, liveness and mediation are seamlessly integrated into everyday life.

For many readers, at first glance, the desire to attend to screens may seem rather counterintuitive. The cultural common sense suggests that embodied, live, or "face-to-face" communication *should* be more (most) relevant to people's lives. In conversations with a wide range of people such as students, family, and friends, when I mention that I am researching digital screens in everyday life, I hear an incredibly consistent anecdote—a brief narrative vignette that has apparently captured many Americans' imaginations. With great disgust and severe judgment, my acquaintances and loved ones describe the scene of a family sitting at a restaurant: While enjoying their food, everyone has their heads down focusing on some portable electronic device rather than talking with one another. Despite this remarkably familiar and recognizable scene, people still seem stunned and appalled that such situations occur. In fact, *based upon the media logic of relevance,* this is what we should probably expect. When fully immersed in the logic of digital screens, activities such as small talk with family members *appear* irrelevant as they are largely "inefficient" and offering satisfaction only via resolute enduring long-term commitments (rather than fleeting pleasure).

Liveness is typically random (inefficient), routine and, by definition, unscreened. Most nonscreened, everyday experiences are, at least by mediatized cultural criteria (fleeting pleasure and efficiency), irrelevant. Something that is screened is, exactly that, *screened* (i.e., filtered). As Auslander (1999) observed: "Currently, with the increased dominance of digital media and the concomitant redefinitions of liveness, mediatized forms still enjoy far more cultural presence and prestige—and profitability—than live forms" (p. 187). As a recent advertising campaign for HBO's mobile App HBOGo demonstrates, after images of bored individuals engaging in inane small talk in spaces such as a city bus or an elevator, the commercial states: "The story you could be watching is better than the one you're in."

Despite a common sense to the contrary, in everyday practice, people appear to have largely embraced the media logic of digital screens—within contemporary American culture, individuals consistently prefer screens to liveness. The privacy settings of Facebook (or the caller ID on a mobile phone) "screen" the communication with members of an individual's social network, the editorial processes of entertainment media ensure that a song or movie is of high entertainment value, and the search function of Google promises that the Webpage is of interest to my immediate concerns. Liveness, as unscreened "content," has no assurances of relevance. In fact, most things that are live are actually quite irrelevant based upon the media logic of digital screens as people are often forced to perceive sensory data that lack value as fleeting pleasure or efficient problem solving. What is on a screen is virtually always *more relevant* than what is live.

Further, these judgments of relevance are less irrational or self-destructive than the cultural common sense sometimes suggests. For instance, any teacher who uses an LCD projector to display content in a classroom knows that students will, almost without exception, attend to *anything* on the screen (unless, of course, something more relevant is on their mobile phone or laptop). As I integrated projected materials into my classes, I quickly learned that if I want students to remember something such as the date that an assignment is due or an essential concept, I should put it on the screen. Students have simply come to realize that the screened information is highly relevant whereas the oral (live) information (often examples, anecdotes, conversational points, and so on) is relatively superfluous. As participants in education institutions that have fully embraced a logic of digital screens, are they wrong?

The now ubiquitous use of digital photography clearly illustrates the pervasive logic of relevance associated with digital screens. Today, virtually all cell phones and other mobile devices (e.g., MP3 players and tablet computers) are equipped with digital cameras and Internet connectivity. Using these devices, individuals can record and, via social networking sites such as Facebook, catalogue what is relevant in their everyday lives. Anyone with young children in the contemporary world knows that parents are now *expected* to record any "relevant" behavior and post/send the photographs to loved ones immediately (for my wife and I, often prompted by pestering grandparents). At restaurants and other public locations, I notice parents busily snapping pictures (or, with more frequency, video recordings) of a baby's random cute smile or toddler's funny gesture that, in the past, would be relegated to memory and oral story telling. A tremendous amount of energy is spent on recording and cataloguing the *routine* behavior of children (not simply "special occasions" such as family vacations, birthdays, or holidays). Further, this is not isolated to parents, with individuals (e.g., college students) also recording and cataloguing the routine, but apparently relevant, details of their everyday lives via the interoperability of camera phones and social networking websites. By recording and distributing routine behavior onto digital screens, individuals' everyday lives *become* relevant to members of their social network (the same behavior, if unscreened, would be deemed relatively irrelevant).

IMPLICATIONS AND CONCLUSIONS

This chapter is an initial outline of the sense-making experiences of individuals navigating a culture immersed in digital screens. As cultural participants, individuals are apparently expected to direct their attention based upon the media logic of relevance. Building upon this initial analysis, fleeting pleasure and immediate (efficient) function appear to be meaningful dimensions of this media logic of digital screens. This contemporary media logic evokes the work of Richard Lanham, Guy Debord, and Jean Baudrillard. I believe by putting my initial observations concerning the media logic of relevance in conversation with these three important theorists, I will be able to uncover a few initial implications.

Richard Lanham (2006) explicates the twenty-first-century economy of attention, an economy built upon the triumph of "fluff" over "stuff." In other words, in the contemporary global economy, information (e.g., media content) has greater value than material goods. Defying the precept of "supply and demand," the more attention media content generates (e.g., the greater number of website visits and circulation of a celebrity's indiscretions) the more valuable the content becomes. Advertisers, television networks, websites, radio stations, employers, loved ones, and politicians all compete in an incredibly competitive marketplace of attention. Building upon Lanham's conclusions, digital screens are arbiters of attention. In contemporary culture, the more something is screened, the more valuable it becomes such as a news story trending on Twitter. In recent years, this has reached a kind of synergy of the absurd in which the eccentric or outrageous behavior of a celebrity (e.g., Charlie Sheen and Kim Kardashian) is so widely screened, it creates its own mini economic system of attention.

The dissemination of news stories via social networking clearly demonstrates the economy of attention associated with digital screens. A Pew Research Center survey found that, increasingly, individuals find their current events news via social networking websites (Anderson and Caumont, 2014). When individuals rely on their digital network (friends, family, favorite celebrities) via social networking sites for current event news, they circumvent institutionalized media gatekeeping by relying on what their favorite Facebook, Instagram, or Twitter users deem as relevant. The more this "viral" media content travels from screen to screen as a trending topic, the more valuable (and relevant) the content becomes. As I explore in Chapter 4, mobile devices simplify the "information overload" of a complex hypermediated environment offering a cognitive map for everyday life.

The centrality of visually stimulating spectacle within the media logic of relevance suggests that contemporary Americans live in a variation of Debord's (1994) "society of the spectacle." Debord described the society of the spectacle as a worldview dominating contemporary culture:

> The spectacle cannot be understood either as a deliberate distortion of the visual world or as a product of the technology of mass dissemination of images. It is far better viewed as a weltanschauung that has been actualized, translated into the material realm—a world view transformed into objective force. (pp. 12–13)

One dimension of the complex, multifaceted worldview of digital screens appears to revolve around the fleeting pleasures of ubiquitous spectacular entertainment. Stated slightly differently, the pleasures and desires of spectacle are part of a larger cultural common sense—desires often taken for granted as screens are obscured from consciousness. In a culture saturated with imagery on digital screens competing for our attention, when spectacular or carnivalesque images of freakish or grotesque sexuality, violence, and the like appear on a screen, eyes are invariably drawn to the images. For instance, the Bush administration used spectacular images of terrorism and militaristic violence to frame foreign policy and steer public discourse after the 9/11 terrorist attacks (Kellner, 2007). More recently, during his 2016 presidential campaign, Donald Trump offered a daily dose of outrageous and incendiary comments that garnered tremendous attention on social media from both his supporters and critics.

This was never more profoundly represented than on a Tuesday morning during my typical morning routine. I brought my cup of coffee to my computer in my office and began reading the online editions of the *Denver Post* and the *New York Times*. On the front page of the online edition of the *New York Times* was an image that haunts me to this day—President Bush appearing on the television show *Deal or No Deal*. *Deal or No Deal*, despite exclusively featuring people choosing briefcases which may or may not offer them cash prizes, was one of most popular television programs of 2008. In other words, as a game show that lacks any discernable competition—for example, no trivia questions or word puzzles—it is a show that is purely the spectacle of prize giveaways amid dazzling imagery and sound (to me, the show has always felt like it is from a kitschy dystopian science fiction movie directed by Paul Verhoeven). After reading the brief article, I found the clip on YouTube. In the broadcast, the host, Howie Mandel, after describing the sacrifice of the soldier who was the game show's contestant, excitedly said, "Look at the Screen!" and pointed at an enormous screen in the center of the set. In a stilted, obviously scripted series of comments, Bush said, "I'm thrilled to be anywhere with high ratings these days" to smattering of laughter from the audience. What haunted me was the relative banality of the event—with a response of slight amusement, the studio audience seemed unfazed and unsurprised by the spectacular nature of the appearance—the most powerful person in the world appearing on a gigantic screen in the middle of a television game show. It fit so perfectly and effortlessly into their contemporary worldview, it barely warranted a response.

For Debord, spectacle is not so much a by-product of technological advancement but rather an ever-present sense-making system that transcends any specific technology or media content. Similarly, it is increasingly apparent that in everyday life, individuals make very few distinctions between so-called separate mediums of communication (mobile devices, televisions, personal computers, and so on), instead relying upon the media logic of relevance to seamlessly move *between* interoperable digital screens (as well embodied live sensory experience). Not only do all digital screens share remarkably comparable technical (digital) characteristics but, more functionally, the spectacle of these digital screens suggests a kind of interchangeability and interoperability. The individual's overloaded attention shifts between these screens based upon (at least in part) the relative spectacularness of the content flittering around the devices.

In the mediatized world of twenty-first-century America, people appear to be enveloped in a world of hyperreality. Baudrillard (1994) famously argued that "Today abstraction is no longer of the map, the double, the mirror, or the concept. Simulation is no longer that of a territory, a referential being, or a substance. It is the generation by models of a real without origin or reality: a hyperreal" (p. 1). In everyday practice, the simulation of reality via screens has become habitually *more* relevant than the "live" version of reality. Mediatization, especially as conceptualized by Baudrillard and Auslander, transforms institutions, social structures, "and larger socio-political process"

(Auslander, 1999, p. 5). From this perspective, mediatization invokes the institutional transformations bubbling up from the media logic of relevance. For instance, my family domestic space has become saturated with digital screens. Domestic spaces are now "an entertainment paradise" with elaborate home entertainment systems (Tryon, 2009, p. 6). In addition, portable digital devices clutter most American homes making their way onto the dining room table and into the bedroom (and, at the risk of being indiscreet, even the bathroom). The family home is now a digital screen environment dominated by the media logic of relevance. Within this logic, family members check each other's Facebook pages while physically sitting together around the kitchen table. Similar observations could easily be made about a wide range of American institutions such as government, education, and even religion. The etymology of the word *screen*—as "skin" (Introna and Ilharco, 2004) further suggests the integration of affective spectacle into individual's bodies, homes, and automobiles (our "skins"). In other words, people are covered in the hyper-real mediatized logic of screens, and screens are another tissue of our bodies. For instance, Burger King formed a promotional partnership with the *Simpson's Movie*. As part of the promotion, fans could go online and "Simpsonize" themselves by uploading a digital photograph and identifying a series of physical features. As a longtime *Simpsons* fan, I Simpsonized my wife and I, and subsequently sent the images to many of my family and friends. Since this successful promotion, a number of similar websites, for instance a website promoting the 2015 film, *The Peanuts Movie* (in this case, "peanutize" someone), have offered the ability to become characters in television and film. The self, reflected back to us on a screen, is "consumed as a sign" (Lefebvre, 2003, p. 102).

As a concluding bookend to this initial analysis, recently, I had the opportunity to attend a National Basketball Association game with my daughter (the same daughter who was watching "Dora the Explorer" in the hotel room). Like virtually all American sports arenas, the stadium featured a large, high-definition screen suspended from the ceiling that displays live images of the game and promotional clips during time outs. Despite the entertaining on-court play (our seats were only a dozen rows from the court), my daughter remained transfixed to the jumbo screen for much of the game. When the dancers began to perform, her eyes gravitated to the screen, not the embodied performers moving below the screen. My daughter had a wonderful time laughing and smiling through much of the game but I left the experience a bit conflicted. Did she miss the "true" experience of the game? Or, was the jumbo screen the *true* experience of the mediatized spectacle of a sporting event? Most hauntingly, I realized that the media logic of relevance is so ubiquitous and overwhelming that even a small child quickly falls into its gravitational pull.

Chapter 3

Postmodern Ethnography of Everyday Life

Julie walks down a busy Chicago street carrying her iPhone, playing a random shuffle of her music collection. She enters a Starbucks, orders a latte, and sits down. The Starbucks plays digital radio over the sound system—a mix of adult modern rock (called The Starbucks Café) featuring Adele, Ray Charles, Norah Jones, and the like. In the corner, a flat screen television (sound turned off) projects CNN Headline News with a constant loop of images of violence in the Middle East, election coverage, highlights from yesterday's college football games, and images of celebrities in turmoil (this week, it is Bill Cosby). Continuing to listen to music on her iPhone, Julie pulls her notebook computer out of her bag and immediately connects into the Wi-Fi network. She checks her e-mail (mostly work messages), updates her Facebook page, and scans the *Chicago Tribune* website. Her phone rings (with a ringtone of her favorite song, "Superstition" by Stevie Wonder) and, when she answers, she hears the voice of her friend, Jack, from Denver.

From a cultural perspective, where is Julie? What is/are her culture(s)? Is it the Starbucks? Is it the virtual community of the Internet? Is it the televisual and auditory popular culture of music, television, and so on? Is it the organizational culture of her job? Is it the relational culture developed with her friend Jack? In general, ethnographers have tended to isolate these cultures studying the virtual community as separate from the coffeehouse culture and examining the popular culture of commercial media as separate from her organizational culture. Yet, considering this increasingly representative example, it is apparent that these diverse, fragmented cultural practices and processes are seamlessly integrated into Julie's everyday life—Julie simultaneously experiences these divergent "cultures" as banal, routine, and ritualistic. For Julie, these are not distinct cultures, rather, these fragments and pieces of everyday life represent a kind of postmodern culture. Further, millions of people experience a similar integration of communication technology and physical environments as they commute to work in their automobiles, relax at home after a stressful day, or enter the hypermediated environments of public spaces like airports, franchise restaurants, and shopping malls.

Technologies so central to everyday life have become invisible objects defying self-reflexive understanding. As noted in Chapter 2, researchers have well documented the astonishing growth of digital communication devices, as wireless technology has "diffused faster than any other communication technology in history" (Castells, Fernandez-Ardevol, Qiu, and Sey 2007, p. 7). The saturation of mediated communication in contemporary cultural life demands that researchers develop alternative methods of social research.

In this chapter, I outline a specific method for understanding hypermediated culture. In addition, I consider ethnography's unique capacity to make sense of these postmodern, fragmented experiences of everyday-mediated life. In order to emphasize ethnographic practice, throughout the chapter, I return to a central question: How do people practically analyze hypermediated cultural experiences? The techniques described can be enlisted by researchers or by individuals simply seeking to make sense of their complex digital social world. I argue that hypermediated cultural life can be navigated and understood via three overarching practices: (1) tune to epiphanies in order to make the invisible visible, (2) maintain self-reflexivity via a perpetual awareness of the contradictory conditions of postmodern culture, and (3) transform bits and pieces of information into coherent (albeit fragmented) cinematic stories.

In order to understand how postmodern, digital experiences fit into everyday cultural practices, as a *postmodern* ethnographer, I embrace the mobilized gaze of the flaneur. After distinguishing between traditional and postmodern ethnography, I describe four principles that can serve as guideposts for the observations of postmodern, hypermediated cultural experiences, leading to more mindful digital communication in the everyday: (1) embrace the emergent, fleeting moment, (2) think dialectically about culture, (3) conceptualize ethnography as self-reflexive, partial, and subjective, and (4) embrace the semiotic tradition. Further, as the reader has likely already noticed, I model the flaneur's reflexive approach by interweaving brief excerpts of popular media (movies, TV shows, and so on) that I encountered during my observations (a technique used effectively by ethnographers such as Vannni, Waskul, Gottschalk, and Rambo, 2010). The pop culture excerpts, which appeared revelatory in the midst of my observations, represent the mediated fragments that fill contemporary everyday life and demonstrate how one might come to reflect on experiences outside the prevailing media logic of digital screens.

TRADITIONAL ETHNOGRAPHIC APPROACHES TO SOCIAL INTERACTION

The detailed observation and thick description of particular communities is at the heart of traditional ethnographic approaches to culture and communication. In other words, historically, ethnography is concerned with "the environment, problems, background, language, rituals, and social relations of a more-or-less bounded and specified group of people" (Van Maanen, 1988, p. 3). In this regard, the communicative understanding of ethnography has been firmly grounded in the anthropological approach represented by Geertz (1983) and the socio-linguistic approach represented by Hymes (1974).

The performance of identification via communication practices, such as ritualistic performances at basketball games (Carbaugh, 1996), has been a primary concern of ethnographers of communication. From this perspective, ethnographic inquiry is "the formulation of the general way in which communication is patterned within a socioculturally situated community, field, or domain" (Carbaugh and Hastings, 1992, p. 161). Defining and/or discovering *coherent boundaries* of cultural identity and community have been central issues in the ethnographic project. These previously coherent boundaries of identity and community have been blurred in the everyday life experiences of the hypermediated era.

As the discussion below articulates, the central problem addressed in this chapter is "How should ethnographers study the experiences of individuals immersed in digital technology?" Grounded in the ethnographic tradition, in my experiences with mobile devices, I sought to make visible mediation processes so ubiquitous that they have become increasingly invisible to cultural participants (see Chapter 2). Deuze (2011) argues that "media become the playground for a search for meaning and belonging" as "people produce themselves (and therefore each other) in media" (p. 138). In other words, in a world in which "everything" is mediated (Livingstone, 2009), the individual is never outside of the "hegemonic signifier" of information (Lash, 2002, p. 199). Drawing upon the work of Lash and Deuze, the problem of hypermediated culture is imperative to contemporary ethnography: How do people understand cultural practices and artifacts that are simultaneously so ubiquitous, interwoven, and enveloping they are largely invisible? In many respects, this project strives to find reflexive light in the vast shadow of technology. Via the practices highlighted in this essay, I believe that ethnography is capable of transforming information hegemony into coherent (yet fragmented) auto-ethnographic narrative or, to paraphrase Deuze's (2011, p. 145) argument, to make our media life into art.

POSTMODERN ETHNOGRAPHY OF EVERYDAY LIFE: BOUNDARY-LESS PLACES AND IDENTITIES

A wide range of scholars have complexly theorized and problematized the concept of everyday life. Specifically, the writings of Henri Lefebvre and Michel de Certeau have inspired explorations of the disciplining of the individual's routine, ritualized daily experiences and the individual's resistance to these constraints on everyday life. As Lefebvre (2003) described the everyday life of the French working class, "there was a power concealed in everyday life's apparent banality, a depth beneath its triviality, *something extraordinary in its very ordinariness*" (p. 37). Further, de Certeau (1984) explored the "tactical resistance" of people in their everyday lives, particularly in their consumption and manipulation of media texts. From this groundbreaking work, scholars have emphasized everyday life (particularly in respect to new technology and commercial media) in terms of the shifting notions of time and space (see Burkitt, 2004; Grossberg, 1992).

A series of profound changes to the time and space of everyday life mark the postmodern era. Postmodern cultures are "characterized first and foremost by mass mediated experiences and new cultural forms of expression" (Harms and Dickens, 1996, p. 211). More specifically, Jameson (1991) described postmodern culture as a world of pastiche "transformed into sheer images of itself" (p. 18). Everyday life is increasingly mediated via technology and digital images dominated by what Baudrillard (1988) referred to as the hyperreality of simulated images. As I asserted in Chapter 2, contemporary posmodern culture is best described as hypermediated or saturated with *screens* projecting spectacular images.

Postmodern theory provides a means of revealing the unique contradictions of hypermediated experiences. It asks a set of questions concerning the inherent dialectical contradictions of hyper-reality, everyday life, mediation, and hybridity. With its focus on economic and social structural realities, postmodern theory draws out both the broad institutional and specific detailed dimensions of hypermediated everyday life. Postmodern theorists such as Jameson and Baudrillard have emphasized cultural changes to social institutions, culture industries, history, militarism, ans so on. As Caron and Caronia (2007) observed "it is in fact through daily exchanges, ephemeral conversations, through the micro-narratives of everyday life, that technologies are connected to emerging dimensions of our social world" (p. 46). The time and space of human communication in a hypermediated postmodern culture are, in many respects, bound-less or boundary-less. For instance, once *inside* the machines of digital media, an individual drifts into the infinity of the World Wide Web (and/or channel surfing) with no end (or boundary) in sight. Further, mobile communication devices are associated with filling "no-where places" and "no-when times" (Caron and Caronia, 2007, p. 98), such as waiting in line or sitting on a train with meaningful, yet fragmented, experiences. As Kien (2009) recognized in the emergence of mobile technologies, "From previous ethnographic perspectives, the field has become unbordered and thus unmanageable" (p. 14).

Gergen (2002) argued that, in part due to the ubiquitous communication technologies of the postmodern condition, identity is no longer "specified" or coherent. Postmodern culture can be understood in terms of spaces *between* bounded cultures and cultural identities—spaces of hybridity. Hybridity is an appropriately malleable and ambiguous concept that has come to connote a kind of cultural mixing (combining of language, art, ritual, and so on), a mixing which is often accelerated by the development of communication technology (Pieterse, 2001).

In many ways, ethnography's historical emphases of detailed daily practices and identity performances offer unique opportunities for understanding the hybridity of postmodern everyday life. As an embodied and sensual practice, "ethnography is the everyday life research strategy par excellence" (Vannini, 2009, p. 5). More specifically, McLaren (1997) argued that postmodern ethnographers "must negotiate the everyday scene of postmodern hybridity" (p. 81). For the present analysis, hybridity is represented in the seamless integration of knowledge work, social networking, entertainment media, and so on in everyday life. As Castells et al. (2007) explained: "the new communication system is characterized by its ability to switch from work to sociability and to entertainment in the same time and space" (p. 110). Sitting in their proverbial cubicles (or in a home office or in an airport), while preparing information work on their computer screens, information age workers interact with a wide range of entertainment-oriented media: they listen to Pandora or Spotify, surf the Internet for celebrity gossip, follow the box score of their favorite sports teams, and so on. Hybridity is also represented in the blurred boundaries of space in postmodern culture: "People are here and there, in multiple heres and theres, in a relentless combination of places" (Castells, et al., 2007, p. 172). Individuals experience many places simultaneously via this "multilocalization" (Caron and Caronia, 2007) of ubiquitous media screens. As cultures are much less-bounded systems, we need forms of inquiry to understand hybridity along the permeable and blurred cultural boundaries of space and time.

In my observations, these strands represent the aforementioned tensions between the power imbedded in popular culture and individual choice in everyday life. Conquergood (1991) eloquently articulated that "instead of dividing lines to be patrolled or transgressed, boundaries are now understood as criss-crossing sites inside the postmodern subject" (p. 184). Boundaries, once clearly demarcated on maps, on street signs, and on our bodies via race and gender, are increasingly discursive, fragmented, and fluid. Stated another way, we are co-constructing moving boundaries moment to moment in the performances of everyday life. Following this line of reasoning, Krizek (2007) argued for a postmodern ethnography that emphasizes "fleeting moments of cultural seepage" and "ephemeral moments" representing "pieces of the mosaic puzzle." For many cultural participants, these moments of everyday life go largely unnoticed as routine and even ritualistic. Nonetheless, these moments are of great significance in postmodern culture—in fact, most of postmodern culture is made up of such moments.

While ethnography has historically emphasized "bounded spaces" or cultures that inhabit a geographic place (e.g., Teamsterville, see Philipsen 1992), in a sense, mobile technology involves carrying our "place" with us, or perhaps more accurately, a perpetual placelessness and/or multi-localization. Further, unlike other ritualized forms of communication such as a basketball game or a business meeting which have relatively clear sequences of action and sustained communication practices (see Carbaugh 1996), the cultural practices associated with the mobile technology are largely an experience of brief fleeting moments without clear unity or sequence.

In recent years, scholars have explored mobile communication technology, examining, in their words, "iPod culture" (Bull, 2007) and "cell phone culture" (Goggin, 2006). I believe one of the fundamental limitations of this recent work, particularly from an ethnographic perspective, is that these authors treat the experience of the cell phone, iPod, or any other digital device as if it was discreet—that is, as if the device is not integrated and experienced with many other screens and fluid/unbounded cultures simultaneously (as explored in Chapter 2). These approaches are understandable, as, historically, media research (influenced by the work of McLuhan) has emphasized *machines* (i.e., the medium of television, radio, the book, or the computer) and ethnographic research has emphasized relatively bounded cultures. While this scholarship has much to offer, particularly from an historical perspective, I cannot help but think that it misses one of the most central features of these experiences with contemporary digital media: the experience of being *between* screens and/or cultures rather than firmly rooted to a single machine or within a cultural boundary. In this respect, postmodern ethnography must "ignore many of the fencelines that serve to organize academic research and writing in the traditional disciplines" (Dorst, 1999, p. 2). The ethnographer's function is to tie together the diverse, fragmented, and disparate threads of everyday life into something resembling coherent and meaningful narratives.

In order to account for these fundamental cultural transformations, scholars have argued for a postmodern ethnography (sometimes called "new" or "post" ethnography). The term *postmodern ethnography* is admittedly (and deliberately) vague, but goes beyond merely "experimental ethnography" and emphasizes the "late capitalist, mass mediated moment" (Borchard, 1998, p. 242). Postmodern ethnography recognizes how contemporary cultural experiences move seamlessly between the physical and the mediated and thus necessarily blends aspects of conventional ethnography (e.g., field notes within a cultural scene) with media (semiotic) criticism. Gottschalk (1995) referred to this process as "a little like channel surfing" (p. 224). Postmodern ethnographers examine "ethnographic fragments" (Borchard, 1998; Gottschalk, 1995), which are often incoherent and contradictory. As a collage-like juxtaposition of postmodern, ethnographic fragments,

> They are not intended to cohere into an integrated story that would claim to represent some sort of final or comprehensive "truth" . . . following various trajectories and arresting moments, [the postmodern ethnographer] produce[s] these subjective fragments by juxtaposing together fleeting interactions, unstructured interviews, postmodern insights, televisual interruptions, and self-reflexions. (Gottschalk 1995, p. 224)

These fragments are what constitute "everyday life" in a hypermediated culture. Further, these are communication processes that defy previous understandings of cultural context, public (mass)/private (interpersonal), embodied identity, collective social action/ritual, and so on. Nonetheless, ethnographers should not fear or lament these changes to cultures; rather, we should embrace these cultural transformations as unique opportunities for ethnographic inquiry. These changes are far from an obstacle for contemporary ethnography—they serve as the juicy stuff.

The Critical Turn: From Micro to Macro (and Back Again)

The postmodern cultural turn demands that ethnographers move between micro practices (the immediately observable activities of everyday life) and macro (general) cultural systems and structures. In postmodern cultures, macro cultural systems and structures take many forms, such as the ideologies of advertising, the software programs (e.g., operating systems such as the iOS) that structure computer-mediated experiences, and the uniform physical architecture of cultural spaces like global restaurant franchises (e.g., Subway or Starbucks). Carbaugh (1996) described a perspective toward communication that "seeks to 'see'—speak, listen, sense, or feel—the 'micro' and 'macro' concerns of identity, from the vantage point of situated communication practices" (p. 11). The "vacillating form" between micro and macro concerns can be understood dialectically (see Carbaugh, 1996). In particular, researchers interested in everyday life must relate "micro-level processes to the macro-level of social order" (Scott, 2009, p. 5) that complexly organize everyday practices. Of course, these societal structures are essentially an immense collection of individual practices and if individuals disrupt these practices, they can disrupt the broader social order. As dialectical tensions, the ethnographer continuously shifts between macro and micro concerns.

Postmodern ethnography must embrace a critical, self-reflexive lens that recognizes the political and ideological implications of our ethnographic practices. As Denzin (2006) stated, "Ethnography is a not an innocent practice. Our research practices are performative, pedagogical, and political. Through our writing and our talk, we enact the worlds we study" (p. 422). Similarly, Madison (2005) argued that the critical ethnographer "takes us beneath surface appearances" by "bringing to light underlying and obscure operations of power and control" (p. 5). In order to reach this goal, scholars have advocated a movement toward personal narrative/biography and self-reflexivity (Bochner and Ellis, 1992), or what is often referred to as auto-ethnography. As Crawford (1996) stated, "Autoethnography epitomizes the reflexive turn of fieldwork for human study by (re)positioning the researcher as an object of inquiry who depicts a site of interest in terms of personal awareness and experience" (p. 167). The researcher embraces and records the self-talk, worries, insecurities, subjective judgments, distractions, and so on experienced as an embodied participant observer. In other words, "Doing fieldwork is a personal experience" (Madison, 2005, p. 8).

In a similar (although in some ways distinct) line of thinking, I also view ethnography as a performance. As Victor Turner (1982) described: "In a sense, every type of cultural performance, including ritual, ceremony, carnival, theatre, and poetry, is explanation and explication of life itself" (p. 13). In this respect, meaning is performed by cultural actors constructing a sort of shared reality "in which 'meaning' is ascribed to the events and parts of experience" (Turner, 1982, p. 18). From this perspective, the seemingly banal routines of everyday life appear more as shared performances representing profound cultural meaning. Turner (1982) argued that performances can "probe a community's weaknesses, call its leaders to account, desacralize its most cherished values and beliefs, portray its characteristic conflicts and suggest remedies for them, and generally take stock of its current situation in the known 'world'" (p. 11). Based on the observations of this book, I came to realize that the everyday navigation *between* digital screens was a form of cultural performance. In diverse cultural contexts such as sporting events, church services, and classrooms, individuals ritualistically shift between screens.

As I surveyed the research related to mobile media, I was disappointed, as Koskinen (2007) observed, that so little theory has been produced regarding mobile technology. Although I value the detailed observations of the banality of everyday life, I quickly discovered that the micro needed to be married to the macro. Shifting my attention toward theorizing, I began to view theory building as a form of performative practice, supported by auto-ethnographic reflections—this is theorizing in situ in the spirit of Benjamin and Barthes.

The Mobilized Gaze and the Ethnographic Flaneur

For ethnographers, the macro and micro dimensions of hypermediated everyday life might be best understood via the mobilized gaze of the flaneur. Typically attributed to the work of Baudelaire and Benjamin, the flaneur's mobile gaze is drawn to "the luster of the crowd" in urban environments (Benjamin, 1968, p. 193). Baudelaire's flaneur is described clearly by Friedberg (1993) as "a paradoxical pleasure: to be at home away from home, in the midst of the world and yet hidden from it, impassioned and yet impartial, here and yet elsewhere" (p. 29). The flaneur is not passive or idle; rather, "the flaneur should/could not merely mingle with the crowd, but is an interactor and thus a constitutor of the people's crowd-like-ness" (Jenks and Neves, 2000, p. 9). In a contemporary context, the flanerie of the flaneur is a form of urban ethnography (Jenks and Neves, 2000) that is uniquely well suited to postmodern culture. Like Bruce Springsteen, the flaneur is most at home in the crowd "out in the street."

The fleeting moments and fragments of postmodern everyday life recall Benjamin's (1999) Arcades Project, which has received a great amount of attention in recent years. Like Benjamin, who is often described as a flaneur, contemporary scholars see great value in recording and interpreting the fragmented, fleeting moments of urbanites—their movement in architecture and city structures, their interaction with mediated forms, and so on (see Jones and Foust, 2008). As Benjamin (1999) stated, "Method of this project: literary montage. I needn't *say* anything. Merely show. I shall purloin no valuables, appropriate no ingenious formulations. But the rags, the refuse—these I will inventory but allow, in the only way possible, to come into their own: by making use of them" (p. 460). Like postmodern culture, the Arcades Project embraced the ambiguous "thrownness" and "unbounded" aspects of experience (Sussman, 2003).

From this perspective, flanerie is both a theoretical concept and a kind of method for understanding the social life of urban contexts (Jenks and Neves, 2000). Further, flaneurs are engaging in the process of "unmooring" ethnography from modernist "epistemological certainty" offering a means to "make way for the creation of new postmodern subjectivities" (McLaren, 1997, p. 78). For instance, sociologists have recognized the value of Benjamin's work for understanding urban, postmodern spaces: "The flaneur introduces a phenomenology of the urban built around the issues of the fragmentation of experience and commodification, opening the way for a micro-sociology of the urban daily life" (Jenks and Neves, 2000, p. 2). Recently, scholars have found the unique positionality of the flaneur as an enlightening explanatory lens for understanding postmodern spaces such as the Las Vegas strip (Wood, 2005, p. 327). Wood concluded:

> Being simultaneously and knowingly aware and asleep, productive and passive, intoxicated and ambivalent: these reflect the paradoxical performance of the flaneur and the post-tourist. This notion of pleasure—playful and ironic—calls into question the power of places and the permanence of our subject-positions within them.

The ethnographic flaneur is a mindful observer of contemporary culture shifting between a careful attention to cultural micro-practices and abstract conceptualizations of macro-structures. Benjamin's (1999) observations offer many compelling examples, such as the astute analysis of a family observing a panorama at a Paris arcade:

> Child with its mother in the panorama. The panorama is presenting the Battle of Sedan. The child finds it all very lovely: "Only, it[']s too bad the sky is so dreary."—"That's what the weather is like in war," answers the mother. . . . The mere narcotizing effect which cosmic forces have on a shallow and brittle personality is attested in the relation of such a person to one of the highest and most genial forces: the weather. (pp. 101–102)

In moving about the contemporary urban environment, the flaneur is *tuned in* to the complex cultural practices as both a participant and a detached observer. In this regard, the flaneur must embrace an inherently contradictory/dialectical perspective, as Coyne (2010) observed, "tuning in involves tuning out" (p. 35). In this respect, the flaneur embraces a deep and profound self-reflexivity. Consider my observations within local coffee shops:

> I have encountered a steady stream of flaneur-like characters in coffeehouses. Their gaze shifted between the "screen world" and the coffeehouse environment. By shifting their gaze between screens and people, they are constantly "moving" but sitting still. For instance, in a crowded coffee shop, patrons removed all the chairs from the other side of the table much like a lunch counter so that a full row of four laptops are set up facing the windows at the front of the coffeehouse.
>
> Each individual stared at their laptops facing the large windows, occasionally glancing out the window to see the people walking by on the busy street. While many people come and go from the coffeehouse, the people (including myself) with laptops are able to remain staring at their screens for literally hours. In addition, a 30-something man stared at his smart phone for several minutes with an uninterrupted gaze, then looked up and glanced at the people

> around him. Outside, sitting at a table, a man talked into his cell phone while his companion stared blankly into space, occasionally nervously looking at her cell phone screen. A teenage girl scanned her cell phone, stared at the screen, head down, slumped over shoulders. As a coffeehouse became crowded, a young man entered with his laptop (business causal clothes) sat next to the electrical outlet, plugged in, and stared at the screen. He talked on his cell phone while staring at his computer screen.

How are we to reflect upon meaning in this seemingly uninterrupted flow of experience? In the examples highlighted in this book, I relied upon the concept of epiphany—flashes of meaning in which academic theory intersected with the details of everyday life. In a practical sense, an ethnographic flaneur discovers moments of revelation and meaning that can be likened to the epiphanies associated with narrative inquiry. Denzin (1992) defines epiphanies in these terms: "rupture routines and lives and provoke radical redefinitions of self" (p. 26). For the purposes of this analysis, an epiphany is a kind of auto-biographical turning point. Within hypermediated culture, epiphanies are inherently cinematic as we are living "in, rather than with, media" (Deuze, 2011). In these moments, it is as if the blurry analogue routine of everyday life is briefly drawn into high definition focus with remarkable clarity. In identifying my brief moments of epiphany, I am not claiming any absolute *capital T* truth concerning hypermediated cultural experiences; instead, I am offering the reader a representative moment of self-reflexive auto-ethnographic meaning making. For instance, pop music, which follows individuals throughout the day in cars, supermarkets, malls, and on their bodies in iPods and cell phones, offers a compelling mechanism (or "soundtrack") for epiphanies. An individual's everyday life comes to mirror the techniques of cinema such as the moment in a movie when the music swells and the movie's star realizes his/her purpose or objective: The Van Morrison song builds with the realization that the friendly nice guy/best friend, not the handsome (and vain) corporate lawyer, is her true love; or the orchestral score grows as the humble computer programmer decides to risk his life to prevent the eminent assassination attempt on the president. In other words, these turning points occur within an internal monologue about everyday existence while immersed in the political economy of pop culture production (particularly ubiquitous media products such as pop music and television shows).

The shifting focus between pop culture and embodied experience is reflected in these observations sitting in a Starbucks:

> The woman in the other corner also wears iPod ear buds and reads a book called "Sex, Drugs, and Cocoa Puffs"—It is strange because she is basically

> my mirror image—a fellow urban flaneur scanning the room, reflecting upon its meaning. I feel uneasy, as though I am revealed or as though I am not particularly unique or special (why should I publish about this experience, why not her or anyone else for that matter?). I return to my reading, find myself drifting into the music of my iPod. I remember how much I love the song Oh Valencia! by the Decemberists when it appears on the shuffle. I lose myself in the song remembering seeing their concert last summer with my wife and nephew. My attention shifts when a man walks in, in his mid-50s, in business casual clothing. He greets the employees, they refer to him by name. He deliberately chooses his spot near the window, carefully sets up his laptop—this is a very ritualistic process, performed not just for himself but for the rest of the people in the coffeehouse. I think of my routine—how I set up my iPod, notebook, etc., manipulating the technology in a very ritualistic process.

As the flaneur is tuned in to these subtle and dynamic cultural practices, ethnographic methods are especially relevant to the hypermediated (or "mediatized") experiences of postmodern culture (Hartmann, 2009, p. 233). In a sense, the flaneur is a bit like Alice through the looking glass, trying to make sense of the puzzling curiosities of a fantastical world. As Friedberg (1993) argued: "The gradual shift into postmodernity is marked . . . by the increased centrality of the mobilized and virtual gaze as a fundamental feature of everyday life" (p. 4). As I explored in Chapter 2, postmodern culture is distinguished by an almost seamless integration of media screens (e.g., computer monitors, televisions, MP3 Players, cell phones, and flat-screen TVs) into everyday life. The flaneur is an astute spectator of these screens. As Benjamin (1999) described: "The flaneur seeks refuge in the crowd. The crowd is the veil through which the familiar city is transformed for the flaneur into phantasmagoria" (p. 21). In a world that, in so many respects, is filled with phantasmagoria, flanerie is the valuable means of ethnographic observation. As a methodology emphasizing an individual's sense-making processes of digital culture, the ethnographic flaneur addresses the need, in Auge's (1995) phrasing, for an "ethnology of solitude" (p. 120).

TOWARD A POSTMODERN ETHNOGRAPHY OF HYPERMEDIATED CULTURE

As this chapter has elaborated, postmodernism presents many challenges and opportunities to further develop ethnography. In this section, I describe four principles that served as guideposts for my observations of postmodern, hypermediated cultural life: (1) embrace the emergent, fleeting moment, (2) think dialectically about culture, (3) conceptualize ethnography as self-reflexive, partial, and subjective, and (4) embrace the semiotic tradition.

Embrace the Emergent, Fleeting Moment

As previously discussed, postmodern culture is marked by a tremendous hybridity of cultural practices and traditions. As many cultural critics have observed, the conventional definitional boundaries of cultures are increasingly artificial (see Kien 2009, p. 113). In an information-age global economy with tremendous migration and ubiquitous entertainment media, can we credibly speak of "Japanese" or "Latina(o)" culture? While these identity markers still have tremendous cultural meaning (such as for political candidates or advertisers), the meaning is increasingly in the blurred boundaries *between* cultures. For instance, Japanese popular culture and technology have been integrated into suburban American teenage lifestyles. Further, unique urban languages have emerged combining elements of Spanish, English, hip hop, text messaging, pop cultural references, and countless other linguistic forms. If ethnographers fix themselves on a particular "culture" or "cultural practice/object" they will not allow the blurred boundaries of postmodern culture to reveal themselves during their observations.

The fleeting and ephemeral mark the cultural hybridity of the contemporary moment. Auge (1995) described a world "where the habitué of supermarkets, slot machines and credit cards communicates wordlessly, through gestures, with an abstract, unmediated commerce; a world surrendered to solitary individuality, to the fleeting, the temporary and ephemeral" (p. 78). These ephemeral moments experienced by individuals represent one of the great challenges of postmodern ethnography—ethnographers never know when these meaningful spontaneous moments will arise (see Dorst, 1989). For instance, in my observations of hypermediated experiences, I have found flurries of novel activity than long periods of tremendous routine redundancy.

Following the tradition of the flaneur, a patient's openness to the unexpected, the baffling, and the counterintuitive (all emerging in the epiphanies described previously) is essential to postmodern ethnographic practice. More specifically, digital technology, despite, or perhaps because of, its essential role in contemporary cultures, is quite disposable, fleeting, and malleable. For instance, in only a few years, smart phones (e.g., the iPhone) designed for web browsing, GPS, and entertainment media replaced cell phones designed primarily for exclusively voice and text-based communication. This is one of Benjamin's greatest contributions to the understanding of material culture—focusing on the "fragmentary, transitory object" (Buck-Morss, 1989, p. 160) as integral to a hybrid, fleeting scene of experience.

Think Dialectically About Culture

Postmodern culture, with all of its ambiguities and diversity, is probably best defined by contradiction and tension (see Dorst, 1999).

Given postmodernism's "multivalent readings," the ethnographer or critic's "goal is not to locate the message but the multiple, frequently conflicting messages" (Blair, Jeppeson, and Pucci, 1991, p. 269). Contradictory and non-unified fragments fill postmodern culture. In fact, postmodern ethnographers should be *primarily* studying these tensions, exploring the vacillation between these contradictory forces. For instance, individuals live in an environment in which large media corporations have tremendous power, and *simultaneously* consumers have greater choice and influence over media content than ever before via social media and extensive streaming services. If a consumer sits in a Starbucks, a corporation shamelessly manipulates the entire environment including the music on the overhead speakers, the images on the video screens, and even the messages on the paper cups. On the other hand, an individual patronizing Starbucks can illegally download music via the Wi-Fi network or mock the Starbucks franchise on her/his social media sites, *while sitting* in a Starbucks coffeehouse. As McLaren (1997) observed,

> postmodern flaneurs/flaneuses negotiate spatial and temporal narratives in private, public, and hybridized spheres and wrestle with the tension between the contingent and the universal, between presence and absence, between utopias and heterotopias, between temporal disjunctions and historical trajectories, and between implosions and explosions of subjectivity. (p. 82)

As Virilio (2004) and Auge (1995) have argued, technology accelerates cultural change—the dialectical movements often occur at a bewildering speed that is barely identifiable to cultural participants. The task of the contemporary ethnographic observer is to illuminate the experience of postmodern culture (via a perpetual awareness of the contradictions of the postmodern condition)—*to slow it down, and bring it into focus*—to allow us to understand this continuous and sometimes rapidly vacillating postmodern experience. The hope is not only for the ethnographer to operate as a more ethical, humane communicator in postmodern hypermediation, but also to inspire the practices of flanerie in others.

As Buck-Morss (1977) highlights, Benjamin's work explored the details of everyday life as they revealed political-economic concerns. Benjamin uncovered economic-political structural conditions via dialectical "associative connection between superstructure and substructure" (p. 156). Ethnographers of hypermediated everyday life face similar dialectical challenges. As the vector of postmodern culture, in my embodied experiences as an auto-ethnographer, I felt tremendous tensions as both a passive consumer of cultural commodities and as a scholar producing academic products. In identifying fragments of postmodern pop culture, complex and contradictory meanings would emerge and evolve during the observations, such as the movie quotations below which I fluidly read as satirical, politically/economically hegemonic, and/or nonsensical:

> Okay. But, I'm saying, it's like, I don't want to cram in sex or guns or car chases, you know . . . or characters, you know, learning profound life lessons or growing or coming to like each other or overcoming obstacles to succeed in the end, you know. I mean . . . the book isn't like that, and life isn't like that (Charlie Kaufman in the movie *Adaptation*)
>
> "I'm expressing my inner anguish through the majesty of song!" (Ron Burgundy in the movie *Anchorman*).

Conceptualize Ethnography as Self-Reflexive, Partial, and Subjective

Postmodern ethnography embraces the inherently partial and subjective meanings discovered and/or created in fieldwork. Ethnographers have long been advocating heightened self-reflexivity (Crawford, 1996), but postmodern cultural experiences require a uniquely complex self-reflexivity. As the ethnographer participates in postmodern culture, (s)he must grapple with her/his own contradictory, incoherent identity performance. As this ambivalent and fragmented identity will certainly be instrumental in the ethnographic process, it must be reflected in one's observations, analysis, and writing. As an ethnographic flaneur, I shifted between being "in the crowd" and detached from the crowd. Due to the speed and complexity of hypermediated experiences, participation in postmodern culture required that I immerse myself in mobile technologies shifting my attention from commercial media, to the identification with the participants in the physical scene, to my identity as a researcher truly from *moment to moment*. Further, I live in hypermediated postmodern culture "24/7"—it is not a cultural context I can easily remove myself from for reflection and recovery. Postmodern culture requires ethnographers to, in Garfinkel's (1967) groundbreaking understanding of ethnomethodology, treat the ordinary as unfamiliar and strange, making the mundane visible when seen for "another first time" (p. 9). In other words, postmodernism requires the ethnographer not only to account for one's "positionality" as a researcher (Madison 2005) but also to *confront* one's complex, contradictory sense of postmodern identity as an *everyday lived experience*.

Everyday life is highly routine and filled with habitual experiences that are remarkably difficult to explicate. Noy (2009) observed that auto-ethnography allows researchers to access "knowledge that might be otherwise inaccessible and undocumentable, including feelings, daydreaming, emotions, and the like" (p. 102). The postmodern ethnographer's task is to cobble together representative bits and pieces of postmodern experience. As Dorst (1989) argued: "If our advanced consumer culture cannot be described as a totality, perhaps, it can be documented piecemeal as an archipelago of lifestyle islands" (p. 2). The ethnographer, from this perspective, is a collage maker or collector (much like Benjamin), cobbling together fragments of postmodern culture. Dorst (1989) continued: "The post-ethnographer's goal ought to be the strategic, critical appropriation of the potential energy of those flashes of radical self-estrangement, moments of epiphany, or jouissance, or vertigo in which one glimpses the dissolution of one's own ostensibly stable subjectivity as ethnographer" (p. 205). Self-reflexivity offers a unique *sensitivity* to the interior subjective dimensions of postmodern everyday life. Early in my observations I realized that I am more than present in these "tales from the field," I am the *sense-making nexus* of the swirling postmodern bric-a-brac that I encounter in my everyday life. My goal is to try to reconstruct fragmented experience via a coherent narrative—fleeting moments and thoughts often involving unpredictable movement and dizzying speed—admittedly, a daunting task.

Embrace the Semiotic Tradition

Postmodern ethnography involves the convergence of two domains—the everyday, routine cultural practices, which have historically been studied ethnographically, and the world of screens, which has historically been studied semiotically. In a sense, this is, *Geertz meets Barthes*. Barthes (1964) was interested in systems of signs and their meanings. For Barthes (1964) the focus of semiotics is "ritual convention" and "public entertainment" (p. 9) in the imagery, sounds, gestures, and other forms of everyday life. These signs take many forms, but semiotic scholarship has emphasized media texts like advertisements, films, and television programs. With an emphasis on "mythology," Barthes (1972) argued that "myth is a language" with power deriving from repetition (p. 20). For instance, Barthes described the "intelligible spectacle" of professional wrestling that articulates "justice" in its distinctions between good and evil. In Barthes' (1977) estimation, meaning lies in "the intertextual in which every text is held" (p. 146), as the text is a "multidimensional space" that is "plural" (p. 146). Stated slightly differently, seemingly anticipating the proliferation of screens and mediated fragments, "the birth of the reader must be at the cost of the death of the Author" (p. 148), as the text "only exists in the movement of a discourse" (p. 157).

Within postmodern culture, our everyday lives are saturated with semiotic codes from entertainment media. In this spirit, Dorst (1989) suggested we (re)define the ethnographic "site" as "an image, an idea, an ideological discourse, an assemblage of texts" (p. 3). For example, in my observations, I visited various sports bars during major televised sporting events (e.g., the Super Bowl, NCAA basketball tournament, and NBA playoffs). In these environments, I encountered countless mediated texts such as television and print advertising, popular music, even ironic references to movies like *Rocky* and *Hoosiers*. As a kind of semiotic loop, or maybe more accurately a semiotic mirror, in a sports bar we often found ourselves watching TV commercials featuring people watching televised sports in a sports bar. All participants in postmodern culture draw upon a vast reservoir of pop culture knowledge to make sense of even a routine set of images. For instance, in the television commercials airing during the 2016 Super Bowl (and discussed endlessly on social media), viewers saw subtle references to celebrities such as Marilyn Monroe, Ryan Reynolds, Jeff Goldblum, and Scott Baio as well as brief allusions to movie/TV characters such as Hulk, Ant-Man, Kung Fu Panda, and George Jefferson (to name a few of dozens of such references). In order to make sense of these images, and the subsequent discussions on social media, the viewer needed to draw upon knowledge of a wide range of movies, TV shows, and so on. Further, the observers of these moments (at the Super Bowl party or the Twitter discussion of the commercials) must contextualize these experiences within this sea of interconnected images. In this respect, the postmodern ethnographer closely resembles "Barthes' mythographer of ideological micro-climates" (Dorst, 1989, p. 206).

While ethnography can learn substantially from the semiotic tradition, semiotic media critics can also learn much from ethnography. As others have argued, but has been too rarely practised by media critics, where and how we interact with media texts is often as (or more) important than what we watch, read, and so on. (see Friedberg, 1993). I want to underscore that it is not a flight of fancy or intellectual exercise to shift between the ethnographic physicality of lived experience and the semiotic imagery of media texts—the contemporary cultural moment involves contextualized participants perpetually shifting between these supposedly distinct domains of everyday life. Media, via mobile devices and ubiquitous screens, is increasingly experienced in diverse contexts, even while in motion (e.g., traveling over 500 miles per hour in a commercial jet) shifting between physical contexts. The cultural practices associated with mobile devices remain largely understudied. The fragments of postmodern cultures allow media consumers to creatively construct a unique of bricolage of mediated texts (de Certeau, 1984). For instance, individual cultural experiences/performances are *shared* via Twitter and Facebook with postings of photographs, videos, and web links as well as commentary about others' media life offering unique forms of cultural shared meaning.

The postmodern cultural participant lives in an image-saturated, hypemediated world. Ott (2004) argued: "Rather than doing a criticism of television shows and films, critics might construct media Texts from various fragments of our semiotic laden landscape. This would, no doubt, be closer to how many readers and audiences *today* actually interact with the endless mediascape" (p. 207). In media-saturated environments, rather than coherent wholes, we experience fragments of media as we swiftly move about everyday life—ringtones, mobile applications, thirty-second commercials, billboard advertisements, channel surfing, YouTube mash-ups, and so on. We experience our identities as a flow between roles, contexts, and structures, moving moment-to-moment from partners to teachers to parents to drivers to men and women, as our devices cross boundaries with/for us. To make any sense of the ocean of media bric-a-brac, in the contemporary world, the semiotic understanding of popular media cannot be separated from the ethnographic understanding of everyday lived practices.

CONCLUSION

In sum, for many of us, hypermediated postmodern culture remains a confusing and contradictory experience. It is unsurprising that so many critics of postmodern culture evoke pessimistic images of alienated individuals living lives of emptiness and despair. Ethnographic observers embracing the nuanced critical lens of the flaneur, however, can provide compelling and evocative analyses of postmodern culture that make sense of the complex cultural moment. The unique ethnographic methods outlined in this chapter offer new directions for ethnographic study. In particular, by enlisting the chapter's overarching recommendations, I was able to better understand the simultaneous meanings constructed in everyday experience with mobile devices. I believe people from varied backgrounds can enlist the methods outlined in this chapter for recording, analyzing, and writing the details of diverse-mediated artifacts/processes that fill twenty-first-century cultures. More generally, for ethnographers from diverse disciplines, this book suggests unique strategies associated with heightened self-reflexivity and the contradictory nature of contemporary everyday life. Further, I believe the following analysis can offer an avenue toward "individual and collective control over reality" (Deuze, 2011, p. 144). In other words, via greater self-reflexivity and attention to moments of epiphany, cultural participants will be able to better construct meaning about hypermediated culture for greater understanding and critique. As we understand postmodern culture further, we might pull ourselves out of the rabbit hole of confusion and pessimism, offering unique opportunities for cultural expression, practice, and ritual.

Chapter 4

iOS as a Cognitive Map within Hyper-mediated Culture

> "'Oh, Kitty! how nice it would be if we could only get through into Looking-Glass House! I'm sure it's got, oh! such beautiful things in it! Let's pretend there's a way of getting through into it, somehow, Kitty. Let's pretend the glass has got all soft like gauze, so that we can get through. Why, it's turning into a sort of mist now, I declare! It'll be easy enough to get through–'. . . And certainly the glass WAS beginning to melt away, just like a bright silvery mist. In another moment Alice was through the glass. . ."
>
> — *Through the Looking Glass*, Lewis Carroll

In the nineteenth century, when Lewis Carroll wrote his indelible stories of Alice's adventures in a mythical wonderland, the phrase "looking glass" was simply a synonym for a "mirror" reflecting the reality around the viewer. Carroll's storytelling transformed the term so that a "looking glass" now metaphorically represents a kind of portal into a new reality in which things are not exactly what they appear. In other words, Carroll drew attention to the fact that mirrors transform reality in many regards—flipping images, creating angles of light and shadow, and framing reality via its borders. Anyone entering a carnival funhouse becomes acutely aware that mirrors distort the viewer's reality. As noted in Chapter 1, the phrase "looking glass" evokes reflective surfaces of an entirely different sort—the mobile glass screens that fill everyday life. Building upon the postmodern ethnographic approach developed in the previous chapter, I turn in Chapter 4 to the Apple iOS as a representative and pervasive manifestation of the looking glass. More particularly, the operating system for the iPhone, iPad, and iPod Touch is a variation of Fredric Jameson's cognitive map. In a postmodern culture, signifiers have accelerated as individuals inhabit an "empire of speed" in which time is radically compressed (Virilio, 2004). Within placeless, media-saturated environments with referent-less symbols and boundless signifiers, the Apple iOS (re)locates individuals as a kind of postmodern compass that limits and filters experience in a world of overwhelming signification. In this regard, the power of the iOS is not what it enables but *what it limits*. Like the edges bordering a canvas or mirror (looking glass), the Apple mobile operating system literally frames an individual's experiences. Building upon the relevance processing described more generally in Chapter 2, an individual's consciousness is narrowly filtered via the specific applications of the iOS. Simultaneously, the untapped potential of iOS applications also opens new possibilities for self-reflexivity via an increasingly aestheticized orientation to everyday life.

The following analysis might be described as an extension of the cultural studies tradition of exploring a single artifact or object in order to understand everyday social practices. Most famously, du Gay, Hall, Janes, Mackay, and Negus (1997) explored the "story" of the Sony Walkman as it offered insight into how "culture works in late-modern societies" (p. 2). More specifically, du Gay et. al believe a single meaningful object offers insight into the "circuit of culture" involving five interrelated cultural processes: representation, identity, production, consumption, and regulation. More recently, Carrington (2012) advocated for "object ethnographies" which "are nested analyses that build a rich analysis of the interaction between artefact [sic] and user in the creation and experience of the everyday" (p. 28). Starting with an essential artifact of everyday life, by charting a social history, the object can "be positioned within broader social and cultural flows or narratives" (p. 28). Ultimately, in the cultural studies tradition, the remainder of this book offers a self-reflexive cultural history of the iOS.

However, my approach to the iOS is complicated by my commitment to considering any device (and the various experiences it helps navigate) in a complex, hypermediated context of everyday life. Unlike cultural formations of the past, contemporary postmodern culture involves myriad digital devices that are intimately integrated into an individual's daily practices. Software such as the iOS cannot be understood simply as a static "text" similar to a public speech or popular film. As Manovich (2013) observed, contemporary digital media should be examined via "the tools, interfaces, assumptions, concepts, and the history of cultural software" (p. 34). In order to understand contemporary "digital culture," researchers must begin with the complex interface between the individual and the cultural software that is woven into everyday experiences.

Further, my approach to the iOS is far from a deterministic view of media and individual consciousness. As Carrington (2012) argued, "Everyday objects embody the values and worldviews of designers and the corporations that employ them; however, in the everyday lives of users they are taken up and redesigned" (p. 28). Mobile devices and cultural software do not determine an individual's everyday experience, nor do individuals behave entirely consciously and mindfully when interacting with these machines. In fact, it is the complex dialectical relationships between cultural software and individual consciousness that is at the nexus of the sensemaking of everyday life. In the present analysis, in addition to charting the history and cultural processes associated with Apple's mobile operating system, I self-reflexively studied the everyday practices associated with my Apple iOS devices (drawing upon the methods described in Chapter 3). For over four years, I analyzed iOS 5, iOS 6, iOS 7, iOS 8, and iOS 9 on both an iPhone and iPad. While inductively analyzing the interactive cultural practices/experiences associated with the iOS on the iPhone and iPad, I explicated the unique pleasures, contradictions, and perils associated with the software of these devices. While interacting with my iPhone and iPad on a daily basis, I came to understand many of the unique everyday experiences and cultural processes (and ultimately important dimensions of the media logic) associated with these devices.

Initially, I should clarify what I mean when I reference Apple's mobile operating system, the "iOS." In general, operating systems administer a computer's basic processes, particularly managing hardware and more specific software functions (e.g., applications). Because of the significant control Apple exercises over its products, the iOS determines more of the experience with these mobile devices than relatively open operating systems for other types of hardware such as personal computers. In their marketing, Apple describes the iOS as the "foundation of iPhone, iPad, and iPod Touch" that is responsible for the "mobile experience" (iOS, 2016). Further, the iOS has a number of "built-in" apps (that cannot be removed) such as messages, mail, safari (web browser), notes, maps, and music that are among the most frequently used (and also establish a design aesthetic for third-party applications). In this regard, the relationship between the iOS and specific applications is quite sophisticated—as the iOS is designed to support (and constrain) specific types of apps. As the iOS includes basic features such as allowing third-party applications via the App Store (first supported on iOS 2) or voice recognition (first supported on iOS 5), new apps are developed utilizing these features. In other words, specific applications are dependent on the parameters and processes of the operating system (e.g., the iOS's camera's integration with social media applications enables efficient photo sharing on Facebook, Twitter, and so on). At a deeper design level, the iOS establishes an aesthetic for the iPhone, iPad, and iPod Touch that is based on simplicity and focus (described further in the next two chapters) and limits the design possibilities for third-party applications. Thus, throughout this book, much like Apple's use of term iOS as the overarching software design of their devices, I refer to the "iOS" as the fundamental characteristics of Apple's mobile software that enables (and limits) particular sorts of applications and, therefore, defines the overall mobile computing experience.

THE CENTRALITY OF APPLE'S iOS IN CONTEMPORARY EVERYDAY LIFE

By all accounts, the story of the devices that run Apple's mobile operating system (iOS) is extraordinary. With the introduction of iOS devices (the iPod, iPhone, and iPad), Apple was transformed from a cast off, also-ran personal computing company on the brink of bankruptcy into the most valuable company in the world. Steve Jobs was transformed from an eccentric, delusional wunderkind who squandered a once promising career to a national icon of innovation and genius (in 2014, the Postal Service announced that they will feature Jobs on a postage stamp). As of 2015, Apple had sold over one billion iOS devices (Reisinger, 2016). Most significantly, consumers consistently report satisfaction levels with Apple's mobile devices substantially higher than the closest competitor (Chang, 2012) suggesting that people (due to lack of a better term) *love* their iPhone, iPod, and/or iPad. When Apple first introduced its mobile operating system or iOS in 2007 on its first iPhone, the reaction from consumers was instantaneous and passionate. Not only were there lines around the block at Apple stores across the country, consumers immediately realized that Apple's first iPhone was only the beginning of something much, much bigger. As Berryman (2011) chronicled, "the iPhone experience elicited desire from users—an almost insatiable desire for more. More of everything" (p. 65). Then, the "something bigger" came into focus as Apple introduced the App store in 2008. Allowing third parties to sell mobile applications (apps) via the Apple App store was nothing short of a sea change in mobile digital communication. The numbers associated with the Apple App Store are staggering, with $20 billion in app sales in 2015 alone (Wakabayashi, 2016).

In addition to the incomparable economic success of Apple's mobile devices, the iOS is also intimately connected to countless cultural experiences and institutions in a wide range of domains of society such as organizational communication, economic transactions, political discourse, education, romantic encounters, and family life. In this regard, Apple's mobile operating system is what Manovich (2013) deems cultural software:

> Search engines, recommendation systems, mapping applications, blog tools, auction tools, instant messaging clients, and, of course, platforms which allow others to write new software—iOS, Android, Facebook, Windows, Linux—are in the center of the global economy, culture, social life, and, increasingly,

> politics. And this "cultural software"—cultural in a sense that it is directly used by hundreds of millions of people and that it carries "atoms" of culture (media and information, as well as human interactions around these media and information)—is only the visible part of a much larger software universe. (p. 7)

Unfortunately, despite the obvious centrality of software in everyday life, largely due to the challenges outlined in Chapter 3, it has been largely under explored across academic disciplines (Manovich, 2013).

The greatest impact of the iOS lies in its tremendous influence upon mobile computing. Utterback (1996) argues that, in communication technologies, forces like consumer preferences, market share, and industry standardization, lead to the emergence of a "dominant design" that defines the form of products across manufacturers and brands. For instance, despite dozens of potential ergonomic forms, the design of typewriters across manufacturers became essentially uniform, as competitors adopted the qwerty design. Similarly, the iOS's form is the dominant design of smart phone operating systems across manufacturers and brands. As Carrington (2012) observed: "The major non-Apple smart phones have evolved to resemble the physical appearance and mimic the performance features of the iPhone" (p. 29). In fact, as Carrington points out, the few competitors that did not conform to the iOS design, such as Blackberry and Palm, saw significant declines in sales. Due to a lower price, Android phones have a larger market share than the iPhone (although when factoring in the iPad and iPod touch, Apple still has over half of the market share) (Hestres, 2013). Nonetheless, even a cursory review of the Android operating system reveals it has adopted the basic design of the iOS.

As a *Wired* magazine article recently argued, "iOS initiated a revolution in design—of graphics and interfaces, sure, but also experiences. The gesture-driven touchscreen changed work, entertainment, transportation, socializing, everything" (Pierce, 2016). In short, the iOS "taught the world to smartphone." The dominant design of mobile devices, all derived from the iOS, are relatively easily observable: "extremely high resolution and the uncluttered nature of the interface design; the intuitive operation of the apps and software" (Carrington, 2012, p. 30). In fact, the design similarities between iOS devices and Android devices are substantial enough that a jury awarded Apple over one billion dollars due to Samsung's infringement of Apple's intellectual property (Wingfield, 2012). Stated simply, to varying degrees, the vast majority of smartphones and other mobile devices (e.g., tablets) are using a variation of the iOS. I realized this when picking up the Android phones of friends and family members. I required no tutorial or explanation. I began swiping away in all the familiar gestures of my Apple devices. Manufacturers such as Samsung know that consumers want the basic interface of the iOS with its prominent app icons, touch screen, and App store, so even with the tremendous risk of lawsuits, they simply reproduce these features at a lower cost.

The astonishing consumer affection for the iOS design is well chronicled. As Lindstrom (2012) observed, "our study subjects loved their iPhones; their brains responded to the sound of the phones the same way they would respond to their boyfriend, girlfriend, niece, nephew, or family pet" (p. 55). On the television show *Portlandia*, when Carrie drops her iPhone, she relives their fondest moments together in a montage aesthetically similar to a movie scene featuring memories of lost love (their first encounter, falling asleep together, texting, finding the answers to trivia questions, and so on). As she imagines the horrifying reality of living without the iPhone, the scene culminates with her weeping alone with a bottle of liquor. For myself, I shamelessly love my iOS devices. When they fail, I feel emotions comparable to heartbreak. Recently, my iPad was reaching the end of its life. It began crashing when running videos and larger gaming apps. I felt betrayed. I eventually replaced it and gave the hand-me-down to my daughter for her preschool gaming apps (more about that later). Exploring my new iPad was evocative of the infatuation of a new romance.

Much like a teenage infatuation, I struggle to say goodbye to my mobile devices by leaving them in my pocket or book bag. In moments when I know I should be fully attentive, such as talking with my wife, meeting with a colleague, or playing with my children, my attention is drawn to my iOS devices. There is an inexplicable comfort in these screens. It is a struggle manifest around me every day while observing couples on dates, coworkers eating lunch, parents at the park with their children, and so on. A revealing moment was at my daughter's school play. Sitting toward the back of the room, I noticed that several parents, when their child left the stage, immediately reached for their phone ignoring the other children's performances (which frankly angered me as my daughter was performing at these moments). As I noted in Chapter 1, the dominant discourse of the culture would call this an addiction, ignoring other cultural dimensions of these habitual behaviors. In this chapter, I explore software design and other cultural patterns associated with mobile applications as important, often overlooked factors in this routine behavior.

The basic design of the iOS (uniform, simply displayed square icons) is now used for home computers, televisions, and even automobile dashboards (Oremus, 2014). In other words, today, all of our screens look an awful lot like the iOS. There is something about this basic design structure (a limited number of prominent squares apps) that seems irresistible to individuals in the contemporary world. For instance, on the television situation comedy, *The Big Bang Theory*, Raj purchases a new iPhone. As his friend hastily removes the packaging, Raj says, "Whoa! What's your hurry, cowboy? Savor the moment." Together, the two men slowly remove the plastic from the iPhone and in unison, with a tone of sexual ecstasy, moan, "Oh, yeah."

Why do individuals love the iOS design with such passion? I believe, as a cognitive map for navigating a postmodern culture, the iOS is a kind of lifeline for surviving the postmodern condition. Norman (2013) persuasively argues that the design of products produces "visceral responses" that range from irritated to satisfying. I believe the iOS involves the visceral response of comfort and security based upon a sense of navigational control. In this regard, it is no surprise someone would clutch an iPhone or iPad much like a shipwreck survivor clutches a life vest. Despite the astonishing popularity and cultural significance of mobile apps, "thus far, we know relatively little about apps as a whole; that is, what kind of technological system they constitute as a cultural platform; and, in particular, what kinds of activities, projects, aims, groups and individuals may access apps—and upon what terms, and subject to what social, and power, relations they may do so" (Goggin, 2011, p. 150). As a means of addressing this tremendous gap in research and theorizing, I explore the unique role Apple's iOS plays in the experiences associated with postmodern culture.

THE POSTMODERN CONDITION

An individual navigating postmodern culture faces acute challenges (often referred to as the "postmodern condition") chronicled by the germinal thinkers in postmodern theory. Postmodern theory is notoriously dense and highly contested. Despite the hazards of entering this terrain, I believe (based upon my observations) this theorizing offers unique insights into the role of mobile devices, particularly iOS devices, in everyday life. Thus, with a bit of hesitancy and humility, I offer three themes of postmodern theory especially relevant to mobile digital devices and everyday life: (1) the acceleration of signs, (2) the fragmented subject, and (3) placelessness. As the themes demonstrate, the postmodern condition unsettles the core aspects of an individual's everyday life—the meaning of symbols, identity, and place/context. The introductory discussion below is not intended to be an exhaustive description of these themes but rather a cursory overview of issues most directly related to the experiences associated with iOS devices. Further, each of these themes is fleshed out further in the analysis later in this chapter.

> Why, sometimes I've believed as many as six impossible things before breakfast.
>
> —Lewis Carroll, *Through the Looking Glass*

Postmodern theory, particularly the work of Jean Baudrillard, has contended that postmodern culture is marked by an acceleration of signs. Jean Baudrillard (1988) famously argued that simulacra, substituting for the real (and therefore unbound by material reality), involve "free floating" signs moving at a breakneck speed about postmodern culture. In a world of simulated reality (a reality which has never existed), artificial sign systems dominate all forms of cultural exchange (notably, economic exchanges built upon advertising) (Allan, 2011). For instance, many individuals seem to prefer the false reality (a fake Paris, New York, or Venice) of places such as Disney World (as Baudrillard notes) or Las Vegas to visiting real places. Trapped with a Marxian "false consciousness," Baudrillard's system of thought leaves subjects with few responses to an endless sea of advertising and popular media images offering an all-encompassing world of hyper-real simulation.

Of course, also famously, a world immersed in simulation produces a kind of meaninglessness in which the subject itself is merely another free-floating sign sailing speedily about hypermediated environments. In the past decade, science fiction and fantasy films have moved from niche genres to smack dab in the center of American movies. My students in particular seem endlessly fascinated by false images represented on digital screens. The most popular films of the past several years such as *The Hunger Games*, *Inception*, *Wall-E*, *Avatar*, *Tropic Thunder*, *The Transformers*, *The Night at the Museum*, and *The Dark Knight* franchise, seem fixated on images of false or alternate realities. The question perplexing individuals is: What do all these images mean (if anything)?

> Who in the world am I? Ah, THAT'S the great puzzle.
>
> —Lewis Carroll, *Alice's Adventures in Wonderland*

The subject's conception of self, which is less tied to their physical body, is fragmented by the postmodern condition. Gergen (2000) famously argues that, in a postmodern culture dominated by technology and hyper-real imagery, identity (as a sense of who we are) is relationally defined and therefore unstable, fragmented, and incoherent. Grossberg (1992) follows this line of reasoning by suggesting that identity in postmodern contexts is largely based on temporary affective investments (e.g., consumption based in advertising) leading to a kind of authentic inauthenticity (selling hyper-real authenticity). For instance, when identity is largely based in consumer lifestyle practices, consumers can primarily find an "authentic" sense of self in the products they purchase. For my family, purchasing locally and sustainably produced food (at Whole Foods) performs an authentic liberal, eco-conscious identity. Identity, even deeply rooted "authentic" identity as represented in the personal histories of musicians such as Bruce Springsteen or Johnny Cash, is merely another commodity fluidly and relationally defined in marketing messages (while shopping, I noticed that Target now sells Johnny Cash t-shirts). In more recent years, "Americana" groups featuring only non-electronic instruments (e.g., banjos, drums, and upright bass) such as Mumford and Sons and the Avett Brothers have sold millions of records and filled arenas with fans (including my wife and I) apparently longing for "authenticity." On the television series *Modern Family*, Phil Dumphy faces a similar crisis in authenticity in a postmodern media-saturated era, often humorously struggling to discover his identity as a contemporary parent: "I'm cool dad, that's my thang. I'm hip, I surf the web, I text. LOL: laugh out loud, OMG: Oh my God, WTF: Why the face?" Extending Grossberg and Gergen's analyses, within social media or other mediated environments, an individual is perpetually constructing an unstable identity on Facebook, Instagram, etc., via his/her own personal (authentic) biography. The question plaguing individuals: Who am I?

> "'And if he left off dreaming about you, where do you suppose you'd be? 'Where I am now, of course,' said Alice . . . Tweedledee retorted contemptuously. 'You'd be nowhere. Why, you're only a sort of thing in his dream!'"
>
> —Lewis Carroll, *Through the Looking Glass*

Finally, postmodern cultures are often described as placeless. According to Auge (1995), physical locations in postmodern cultures have no clear sense of place or time. Postmodern non-places (e.g., airports and shopping malls) cannot be defined as relational, historical, or via identity largely due to the compression of time and place (accelerated speed/movement) and an enveloping "world of consumption" (Auge, 1995, p. 106). Ultimately, non-places' empty individuality as regional differences is erased. For instance, Wood (2009) describes an *Omnitopia* as a "structural and perceptual enclave whose apparently distinct locales convey inhabitants to a singular place" (p. 3). For individuals, omnitopic non-places such as sprawling shopping centers, airports, or tourist spaces (e.g., the Las Vegas strip) are highly disorienting, as the embodied subject exists in spaces that are ahistorical and indistinct. The vexing question for an individual is, where am I?

Postmodern thinkers are often (fairly) accused of constructing a theoretical system that traps the subject in a nihilistically hopeless world with no possibility for escape or social change (Slocombe, 2013). Is there any way out of this meaningless, fragmented, placeless postmodern world? Here, cultural theorist Frederic Jameson offers a compelling glimmer of hope—a glimmer that might help explain the tremendous popularity and centrality of mobile devices (particularly iOS devices) in contemporary culture.

Apple's IOS as Cognitive Map

Jameson offers the notion of a "cognitive map" to describe a new art aesthetic to escape global late capitalism. Jameson (1991) describes the cognitive map as: "Surely this is exactly what the cognitive map is called upon to do in the narrower framework of daily life in the physical city: to enable a situational representation on the part of the individual subject to that vaster and properly unrepresentable totality which is the ensemble of society's structures as a whole" (p. 51). In postmodern cultures, an individual is set adrift in a world without clear spatial, temporal, or informational relationships. As Ott and Lazic (2013) describe: "For Jameson, the postmodern subject is overrun and overwhelmed by an endless flow of signs, images, and information, producing a decentered, fragmented, schizoid subject" (p. 262). Virilio (2002) describes the radical transformation of the time-space relationship in terms of temporal compression:

> The escalation to the maximum speed of interactivity, modifies the cognitive map. Every time you have an orientation in space, you have a relation of space and time. Does it take me six months to go to China, or six hours? Will it take me two seconds, or two hours? Thus temporal compression acts on the body proper through the intermediary of the world proper which has been modified in the consciousness of the individual. (p. 76)

According to Ott and Lazic (2013), the compressed time and space of everyday life are ubiquitous and overwhelming for the postmodern subject: "The current space of multinational (global) networks abandons the spatial figures of the grid (city) and territory (nation-state), creating a postmodern hyperspace of total immersion" (p. 262). In general, the individual can no longer locate his/her subjectivity in hyperspace.

Jameson (1991) believes cognitive maps can reorient an individual in the disorientation of the postmodern condition: "An aesthetic of cognitive mapping—a pedagogical political culture which seeks to endow the individual subject with some new heightened sense of its place in the global system" (p. 54). As the term suggests, cognitive maps relocate an individual spatially in postmodern environments by "our positioning as individual and collective subjects," which allows individuals to "regain a capacity to act and struggle which is at present neutralised by our spatial as well as our social confusion" (p. 54). The potential "invention and projection of a global cognitive mapping" offers the possibility to reposition the subject via a new aesthetic. For Jameson, cognitive maps should recenter the people for the purpose of resistance (Ott and Lazic, 2013).

The standardized and popular apps of the iOS function as a kind of apolitical cognitive map for individuals immersed in a postmodern hyperspace. Although the iOS can be personalized in myriad ways, many of the same apps appear on most iPhones and iPads. For the sake of clarity, I will focus on the standard iOS apps and the most downloaded apps in the Apple App Store. The standard apps (apps that appear "out of the box" for iPhones and iPads) on iOS devices include phone (for making phone calls and receiving voice mail), mail (for e-mail), safari (for web browsing), music (for listening to downloaded music), maps, messages (for text messaging), calendar, camera, photos, clock, weather, game center (for video games), notes, reminders, iTunes store (for purchasing music and movies), news, compass, and settings. The standard home screen has remained largely the same (with minor aesthetic changes and the inclusion/exclusion of a few apps) since iOS 2. Further, the most downloaded third-party apps from Apple's App Store between 2012 and 2014 included YouTube, Instagram, Facebook, Pandora, Temple Run, Pinterest, Twitter, Skype, Candy Crush Saga, Vine, Google Maps, Draw Something, Flashlight, Snapchat, and Despicable Me: Minion Rush (Graham, 2013; Lynley, 2012). I first explore the routine experiences associated with these standardized and extremely popular apps. I later consider the possibilities for personalization and the potential performance of resistance in the conclusion.

In general, via the "softwarization of culture," "software has become our interface to the world, to others, to our memory and our imagination" (Manovich, 2013, p. 2). As a cognitive map for navigating postmodern culture, the iOS answers the questions vexing the postmodern subject: Who am I? Where Am I? What do all these images mean (if anything)?

> "'Would you tell me, please, which way I ought to go from here?'
> 'That depends a good deal on where you want to get to,' said the Cat.
> 'I don't much care where—' said Alice.
> 'Then it doesn't matter which way you go,' said the Cat."
>
> —Lewis Carroll, *Alice's Adventures in Wonderland*

Where Am I?

As Jameson (1991) suggests, cognitive maps provide the subject with a "heightened sense of place in the global system" (p. 54). Via global positioning systems, mobile applications such Google maps and Compass literally locate the individual (on Google maps, the individual holding the phone is represented by a soft blue circle) in the postmodern (omnitopia) city-scape. For myself, this is most pronounced when in a strange city, traveling in the disorienting sprawl of urban environments such as Los Angeles or Las Vegas. Global positioning or "location aware" apps allow users to focus their attention, to avoid the disorientation associated with cluttered (with visual signifiers) placeless spaces. For instance, Google maps allow the user to only view restaurants (or shoe stores or coffee shops, or whatever else the user selects), and focus (filter) their attention on "relevant" information in the environment. Importantly, when looking at their digital screens, subjects no longer see the entirety of the physical context—they only see the carefully screened relevant data (returning to the themes from Chapter 2). At a conference in Washington, D.C., a city with which I am pretty unfamiliar, I consistently placed my walking destination in my iPhone and, while walking with my headphones in my ears, the Google maps app gave me audio prompts (e.g., "turn left at 11th Street in 20 feet"). I simply followed these instructions, quite oblivious to the physical cues and image-saturated environment around me. Similarly, whenever I drive to any unfamiliar location in my own city of Denver, I immediately place the address in my Google maps app on my iPhone and follow the directions, often without orienting myself to the material reality around me. When my consciousness is immersed in the mobile map application, I only see myself (represented with the pleasing blue icon) and my destination—the rest of the cluttered and confusing postmodern landscape is filtered out and invisible.

Movement through space is increasingly informed by social networking in which apps like Yelp, Foursquare, Twitter, and Facebook help people describe and recommend places within postmodern omnitopias such as restaurants, bars, and clothing stores. Placeless postmodern environments, when defined by the routine presence of members of my social network, offer greater meaning and clarity when tied to the shared points of identification (e.g., I like the same coffee shop and burrito stand as my friends Dave and Allison). For instance, my wife and I choose new restaurants (such as the enormously popular local food and environmentally friendly Denver bistro, Root Down) largely based on the recommendations offered on social media apps such as Yelp comforted that other liberal, middle-class foodies also enjoy this dining experience. Social media apps demonstrate how space (where I am) and identity (who I am) are intimately interconnected, as is further explored in the following section. The important point here is how seamlessly the iOS, particularly in the form of location aware apps, integrates the self in the fragmented postmodern cityscape.

Similarly, weather apps clarify and simplify my orientation to my physical environment. The Weather Channel app, via location awareness, provides immediate information about the temperature, forecast, and any inclement weather alerts. In other words, I do not need to gaze at the sky or even feel the temperature of the air on my skin in order to glean clues about present or future weather conditions. The relevant climate information (particularly related to the postmodern experiences such as fashion/clothing and travel) is mediated—*screened*—by the weather app.

Spatial relationships are intimately tied to temporal relationships. As the subject moves through space, the speed and rhythm of movement provide meaning to physical locations. For instance, music can define space by providing a sort of soundtrack to physical movement (Bull, 2000). In addition, "no-where" places (Caron and Caronia, 2007) such as waiting in line or sitting on a bus can be transformed by playing a simple video game such as Candy Crush (as I explore below, the most popular games are typically based on simple rules) relocating the individual via the binary code of game play. As the next two sections explore, this relocating of the subject is also tied to broader sense-making processes as the iOS helps position the individual in the dizzying universe of postmodern signs. Stated slightly differently, locating oneself in a postmodern context inherently involves defining one's identity (e.g., defining my identity as an American or as an urbanite *locates* me in the United States or in the city of Denver, Colorado) and hyper-real simulation (e.g., American-ness or Colorado-ness as constituted in digital imagery).

> I know who I WAS when I got up this morning, but I think I must have been changed several times since then
>
> —Lewis Carroll, *Alice's Adventures in Wonderland*

Who Am I?

An astonishing number of apps functionally confirm the lifestyle practices of postmodern subjects. Digital photograph and video sharing are central to the experience of the iOS via popular apps such as Facebook, YouTube, Vine, Twitter, Pinterest, and Instagram. In fact, the camera app of the iOS allows (even encourages) me to instantaneously (i.e., one "click") send photographs to social networking sites or via e-mail and text message. Like millions of others, whenever my wife and I have an interesting experience, such as attending a concert of one our favorite bands (e.g., the aforementioned Avett Brothers) or visiting a national park, we stage photographs explicitly designed to be shared with our social network. Upon viewing these photographs on social media, others choose to "like" the picture (or perhaps ignore the picture) and/or make comments about the picture, confirming or disconfirming the subject's performance of identity. In the spirit of classic psychological theories such as the looking glass self and symbolic interactionism, the subject can explicitly see others' perceptions of his/her identity performances. In a sense, individuals are photographically performing, "this is who I think I am" and others are responding, "Yes! This is who you are."

Parenting, a source of significant uncertainty in the present culture, is one of the clearest examples of performing identity-related practices via mobile devices. In the past, parents largely performed the institutionally prescribed roles of mother and father. However, in the past twenty years, parenting has become a kind of knowledge work in which parents, based on expert advice in parenting books, discussion forums, and blogs, choose a parenting style (e.g., attachment parenting) (for a history of parenting, see Stearns, 2003).

As a new parent, my wife and I felt tremendous uncertainty, feeling pressure to choose the correct practices associated with toys, activities, classes (yoga, gymnastics, and so on), and the like. With a network of other new parents as well as our extended family and friends on Facebook and available via text messaging, our photographs of everyday practices invited "likes" from others, often immediately confirming that we were "good parents" in a culture without prescribed institutionalized parenting roles. Within even a few minutes at any children's museum or sporting event, I notice parents "staging" photographs with their children to be submitted immediately to their social network. So in a postmodern landscape, in which no institution retains authority on defining identity (father) or cultural practices (parenting), the iOS's apps help filter and frame the experience.

This becomes even more explicit as subjects share their lifestyle performances seeking affirmation from others. Much of the postings on sites such as Facebook, Twitter, and Pinterest are the daily lifestyle practices associated with postmodern culture. As a consumer, I must choose from an almost infinite menu of music to stream/download, food to consume, clothes and hairstyles to don, political causes to support, and so on. Of course, any time spent on social networking reveals a wide-ranging discourse concerning these lifestyle practices (e.g., "liking" favorite restaurants, songs, movies, and clothing designers). As Grossberg (1992) observed, Americans feel as though "maintaining a lifestyle was a full-time job which absorbed all of people's energy and time" (p. 303). Via iOS apps, I can confirm my lifestyle choices, which can seem overwhelming in the fragmented world of postmodern culture. For instance, music has long been associated with cultural identity performances (country music's conservatism, alternative rock's outsider persona and so on). By following peers' music choices on Pandora, iTunes, Twitter, or Facebook, an individual can rest assured that he/she is performing his/her cultural identity appropriately. Specifically, iTunes and Pandora's software algorithms limit music choices, automatically choosing music for the user. I feel a tremendous relief at simply pressing play on iTunes or Pandora, released from the responsibility of choosing a song or artist. When I choose the Jack White station in Pandora, I am immediately relieved of the burden of choosing music with the appropriate level of cultural hipster cache or critically acclaimed quality (both important dimensions of my postmodern sense of self). In other words, these apps offer a cognitive map to culturally recenter my fragmented identity.

> Curiouser and curiouser! cried Alice.
>
> —Lewis Carroll, *Alice's Adventures in Wonderland*

What Do All These Images Mean (if anything)?

In a postmodern world that features an explosion of hyper-real signifiers, a kind of meaninglessness and cynicism can overcome the individual. By limiting an individual's field of vision and narrowing possible interpretations of images, the iOS simplifies the boundless image system swirling about postmodern culture. Apple (in)famously exercises tremendous control over its devices, operating systems, and apps (reviewing all apps before they can be sold in the App Store) receiving criticism from consumer advocates, software designers, and others (see Goggin, 2011). While Apple's draconian control (creating an ostensibly "closed" system) has compelled some users to abandon their Apple devices for the more open-source and flexible Android devices (or "jail break" Apple devices in order to use open-source apps), this closeness and control is also an appealing feature.

For individuals overwhelmed by an infinite stream of images in a postmodern culture, the simplicity and limitations of Apple's devices are actually quite attractive. For instance, in a hypermediated culture, the sheer volume of "news" sources reaches dizzying proportions paralyzing an individual overwhelmed with options. In 2015, Apple released its "news" app for the iOS (Greenberg, 2015). The application filters news stories based upon the user's preferences. For instance, for my news app, I selected to feature news about specific topics such as U.S. politics, NBA basketball, foreign policy, and rock music and selected to feature stories from channels such as the *New York Times*, *National Public Radio*, *Entertainment Weekly*, and *Wired.* More importantly, the app collects information and features stories similar to the news the user has read (e.g., "clicked on") in the past. I have noticed that the more I use the app, the more it anticipates the news I will read. Of course, I know I *should* read stories about the suffering of refugees in war-torn and poverty-stricken countries, but during a stressful day, I *actually* choose stories about the latest science fiction movies and music reviews. Eventually, the stories about refugees are filtered out and the entertainment media stories are increasingly featured. The app has cognitively mapped the postmodern culture, without the shame of consciously skipping disconfirming information, exclusively revealing pleasing information that confirms my postmodern identity as a white, well-educated, middle-class man.

The phenomenon of the second screen also demonstrates an individual's postmodern sensemaking via the iOS. Particularly, while consuming television, many people now often have a phone or tablet computer on their lap following live comments on social media. Sporting events such as the Super Bowl or the NCAA men's basketball tournament and reality television shows such as *The Bachelor* or *The Voice* involve a particularly large amount of second-screen usage (and account for an explosion of activity on sites such as Twitter). In 2014, more than a billion tweets were TV related, typically stating an opinion of the content of the television program (Ingram, 2015). Simple tweets provide a moment of confirming sensemaking for a favorite performance or perceived miscue. This is particularly essential for complicated events involving racial codes (e.g., the "blackout" of the Academy Awards due to the underrepresentation of non-white actors), sexism (e.g., the domestic abuse of professional athletes), and so on. As I explore at the end of this chapter, the echo chamber of confirming like-minded commentary further simplifies these complicated political discourses.

Even the simple act of choosing a TV show or movie, once limited to three channels or two or three local movie theaters, now requires sifting through thousands of options on streaming services such as iTunes, Netflix, and Hulu. Further simplifying my postmodern sensemaking, the movie and television review website Rotten Tomatoes offers a simple percentage score (e.g., the movie *Spotlight* has 96 percent positive reviews from critics) for a movie or TV show telling me immediately (in the form of a single number) if a movie is good or bad. Similarly, the most popular media-streaming sites such as iTunes and Netflix also list a simple, single numerical score (based on user reviews) for movies, TV shows, and music. Much like Apple News, iTunes and Netflix recommend media content based on my previous choices providing a significantly simplified decision-making process.

Further simplifying the complexities of the hypermediated culture, the most popular game apps such as Candy Crush (and previously popular games such as Angry Birds) reduce the subject's attention to simple "this or this" binary code relationships. The most downloaded games of all time on Apple's App Store are Candy Crush, Angry Birds, Subway Surfers, Despicable Me, Clash of Clans, Temple Run, Angry Birds Rio, Temple Run 2, and Words with Friends (Moscaritolo, 2015). Reviewing this list of the most popular games, the generic characteristics are relatively clear. With the exception of Clash of Clans, the other games require little mental investment on the part of the gamer. The player moves through the game by solving simplistic binary problems. Generally based upon simple puzzle designs (similar to classic games such as *Tetris* or Mario Brothers), the user chooses from a highly limited set of movements such as up/down, left/right or jump to reach a basic objective. This is in stark contrast to the most popular PlayStation 4 games such as *Assassin's Creed Syndicate*, *Uncharted*, and *Grand Theft Auto 5*, which involve complex narratives and sophisticated forms of decision-making. In other words, mobile gaming within the logic of the iOS is quite different from other forms of video gaming. In moments of stress, these simple games provide me a sense of calm (apparently also experienced by my daughter, as described in Chapter 5). I notice this most clearly in moments of anxiety, such as waiting for a plane to board and takeoff. Looking around airports, I see large numbers of nervous flyers busily playing games such as Candy Crush and Angry Birds. Further framing the individual's experience, iOS games (and more generally, the game-like features of iOS apps) are explored in greater depth in Chapter 5.

At first glance, the repetitive routines associated with the iOS such as selecting a music playlist, reading news reports, or playing simple video games may appear insignificant and banal. As stated previously, everyday life is best defined as meaningful repetitiveness. In this regard, the repeated routine activities that fill daily experiences, such as apps used multiple times each day, come to define the values and meanings of everyday life.

CONCLUSIONS

In 1931, Alfred Korzybski famously aphoristically argued: "the map is not the territory." His point was elegantly simple—via models (often in the form of symbols), humans frame, distort, and limit their perceptions of material reality. The relationship between symbol use and perceptions is well-trod ground raising a wide range of issues such as the complex interrelationships between culture, perception, and language (e.g., Sapir-Whorf hypothesis). More recently, scholars have studied the "public screen" which, on a mass communication scale, frame perceptions of political events via the dissemination of spectacular images (see DeLuca and Peeples, 2002). I believe similar processes occur via mobile devices that serve framing functions in my everyday life. By offering simplified answers to the complicated and confusing questions of the postmodern condition, the iOS is a cognitive map for navigating a world in which images are overwhelming and meaningless, identity is groundless and fragmented, and cultural spaces are illusory and indistinct. An important dimension of the experiences associated with the apps of the iOS involves situating the postmodern subject by framing, filtering, and limiting the field of sensory data.

During the 1980s and early 1990s, probably popularized by Howard Rheingold's influential book *The Virtual Community*, the term "virtual" started to enter the popular vocabulary associated with technology and has persisted ever since. The term is typically used to describe mediation of any technology as the polar opposite of the embodied or real (e.g., a virtual communities or virtual organizations appear exclusively on the Internet). In many regards, Carroll's looking glass was a precursor to this polarization. Baudrillard's distinctions between the real and the virtual (or hyper-real) certainly extended this reasoning even further. Incidentally, the movie *The Matrix* borrows from both Lewis Carroll and Jean Baudrillard (with explicit references to both) recognizing these similarities. According to this reasoning, much like with *Alice in Wonderland* (or Neo in *The Matrix* if you prefer), whatever is on the screen or crosses the threshold of the looking glass, is *unreal*. Today, this simplistic bifurcated view of mediation is conceptually dissatisfying on many levels. As Fernback (2007) argues: "The distinction between the 'real' and the 'virtual' has become much less useful as the internet is firmly ingrained in daily cultural existence" (p. 53). As cognitive maps demonstrate, the "embodied" and the "mediated" intersect consistently as individuals use screens to make sense of the experiences of everyday life. Importantly, on a practical level, what appears on digital screens has very *real consequences*—as job applicants discover when employers comb their Facebook pages, as celebrities have realized after posting racially insensitive tweets, or as politicians have discovered after sending sexually explicit texts. Further, no one would suggest that a soldier in Iraq experiencing the birth of his baby in North Carolina over Skype is witnessing a false, or unreal, event.

Yet, the idea that "virtual" communication is fake or unreal continues to pervade the common sense associated with digital technology (even as people live their lives via these devices). As almost all of everyday life appears on screens (as described in Chapter 2), this screened experience is, for lack of a better term, our *only* reality. This real-virtual distinction has simply past its expiration date for a culture dominated by screened mediation. In fact, the cognitive mapping function of the iOS indicates the complex ways that screen process and filter reality for the individuals. I believe, as I explore in greater depth in Chapter 6, individuals should instead focus on whether or not experiences offer more meaningful and ethical dimensions to their lives (whether or not it is on a screen seems rather unimportant to people navigating a hypermediated culture).

Recognizing the cognitive mapping function of the iOS opens unique opportunities for new discourses. Although the design aesthetics of the iOS functions as a kind of cognitive map for the postmodern subject, it lacks a central feature of Jameson's conceptualization: resistance to global capitalism. As I explore in Chapter 6, iOS applications have the potential to cultivate self-reflexivity via an increasingly aestheticized orientation to everyday life. Unfortunately, the cognitive mapping processes of the iOS are far from politically resistant, and promote remarkably little self-reflexivity. The experience of most iOS users (myself included) is largely consumers dutifully following corporate cues (purchasing iTunes music, following fashion trends of celebrities on Twitter, commenting on the latest Hollywood blockbuster, and so on). Mobile apps are profitable primarily from gathering consumers' personal data (Getting to Know You, 2014) or performing free intellectual labor in the form of product reviews. Unsurprisingly, advertisers are gearing up for a day to collect endless data on individual consumer's behavior and simply prompt them (via their mobile apps) into purchasing products (some might argue this day has already arrived). In this regard, mobile apps largely provide a cognitive map framing the purpose of everyday life as obedient consumption.

Similarly, news apps (such as Apple's news app) increasingly involve filtering processes, so individuals only receive information featuring like-minded commentary. Social media apps, featuring self-selected news sources to "follow," have similar filtering processes. For instance, retweeting on Twitter seems to follow a community structure that is ideologically segregated and highly polarizing creating a kind of echo chamber of homogenous political opinions (Conover, Ratkiewicz, Francisco, Goncalves, Menczer, and Flammini, 2011). With alternative messages filtered out, one's political leanings may become normalized and other perspectives perceived as ill-informed or deeply misguided. Since the emergence of mobile applications, I have noticed a discernible shift in conversations across political ideologies, particularly in my extended family. Conservatives talk judgmentally about liberals as ignorant and naive while liberals speak critically of conservatives as unfeeling and greedy with a kind of taken-for-grantedness that I simply don't recall before these devices were ubiquitous. Perhaps more revealingly, I feel a tremendous sense of self-satisfaction and belonging (a feeling that *of course* we're totally right and they're totally wrong) when a confirming political tweet or item on the *Huffington Post* crosses my iPhone or iPad.

Nonetheless, as I explore further in Chapter 6, the flexibility of the iOS also offers the potential for a profoundly resistive discourse via the performative aesthetics of everyday life. In the process, simplistic, habitual beliefs can be disrupted opening opportunities for more complex and nuanced understandings of the world. For instance, while Instagram and Facebook are frequently used to post banal images (e.g., pictures of one's breakfast at Starbucks), they could more consistently call into question the routines of everyday life by performing alternative perspectives. Rather than allowing corporate marketing to dominate the cognitive mapping of everyday life, via social media, individuals can collectively form a new aesthetics for navigating postmodern culture. Scholars should both discover and help create new enclaves via iOS apps that frame and focus the postmodern subject's experience around an aestheticized resistance. Together, postmodern subjects can modify and personalize the iOS to create cognitive maps that resist contemporary hyper-real placelessness by (in Jefferson Airplane's famous words inspired by Lewis Carroll) "feeding their heads" with alternative and resistant discourses.

Chapter 5

Apps as Gamification in Everyday Life

Since my daughter was about three years old, she has *loved* to play on my iPad. In fact, like most of her friends, she typically prefers to play on the iPad more than any other activity, more so than her crayons/markers, dolls, books, and movies. She loves the iPad so much that we have a series of limitations involving how long and when she can use the device—including setting the timer on my iPhone with a loud alarm indicating her thirty-minute allotment is over. When she uses the iPad, she truly "plays" on the device as she shifts between games and apps effortlessly—moving between her favorite games (cooking games involving making cupcakes, ice cream, and so on), sending text messages to her grandparents, checking the weather, playing videos (on the PBS kids app), taking and scrolling through photographs, and many other activities. Observing her, she clearly loses track of time and space when in the iPad environment. She appears, for lack of a better term, meditative—in state of calm focus. In other words, for my daughter, the iPad induces a flow-like state.

In this chapter, as a continuation of the analysis of iOS devices from Chapter 4, I argue that game-like or gamified apps are particularly well suited to the media logic of relevance by framing, focusing, and filtering the complexities of a media-saturated culture via clear rules and simple numerical outcomes. I reveal the experiences associated with the iOS software that evoke a distinctive emotional relationship between individuals and mobile devices—specifically, I argue that a sensual pleasure is associated with the game-like routine practices of the Apple iOS. The pleasure emerges from the intersection of play, everyday ritual, control, and aesthetics that creates a flow-like experience for the individual. As a gamified experience, the routine practices of hypermediated everyday life are further simplified and clarified and ultimately transformed into the pleasures of video gaming. Enlisting the methods and procedures described in Chapters 3 and 4, the following analysis describes the gamified practices associated with Apple's mobile devices as a source of (quasi)transcendent pleasure via everyday performances—a perverse intersection of sacred ritual forms and consumer culture.

CULTURAL SOFTWARE

As discussed in Chapter 4, everyday life in the contemporary culture is dominated by software. In order to understand the role of digital technology in everyday life, researchers should examine the unique characteristics and practices associated with "cultural software" (Manovich, 2013, p. 34). As Manovich (2013) argues, the software interface changes all dimensions or "properties" of a mediated object, phenomenon, or experience (e.g., a photograph changes when viewed in a web browser, Photoshop, or Adobe Acrobat and a social interaction is altered when experienced via Facebook, Skype, or text messaging). In this chapter, I am studying the iOS as a software interface. Manovich suggests that the software medium is "defined by application software and experienced by users," where "a 'medium' is a paring of a particular data structure and the algorithms for creation, editing, and viewing the context stored in this structure" (p. 336). As Manovich observes: "While 'media effects,' 'media representation,' 'media industry,' 'media theory' and 'media history' have been extensively discussed in large number of books and articles in a number of academic disciplines, this literature does not usually include analysis of software tools and platforms" (p. 340). This book is an initial step in filling the void in the study of software as a medium of communication. In order to understand "digital culture," scholars should survey the complex interface between the individual and the cultural software that is woven into everyday experiences. Specifically, as daily practice, using iOS software is often a game-like experience.

Gamification

Gamification typically refers to "the use of game design elements in non-game contexts" (Deterding, Dixon, Khaled, and Lennart, 2011, p. 2). At its root, gamification emphasizes the dimensions, characteristics, and implications of games in contemporary culture. From this perspective, games are "specialized systems that are characterized by explicit rules and the attempt within those systems to achieve certain goals or outcomes" (Giannetto, Chao, and Fontana, 2013, p. 197). More specifically, gamification studies often emphasize "elements" of games in cultural practices such as incentives and feedback loops. For instance, Foursquare represents gamified social media, providing users methods "for charting their progress, comparing themselves to other users, [and] providing positive feedback when certain milestones are reached" (p. 199). In this regard, gamification is closely aligned with the interests of game studies, a burgeoning field emphasizing (among other things) the enormous popularity of video games in contemporary cultural life.

The term gamification has been applied primarily in two fields: marketing and education/training. For example, Knapp, Blair, and Mesch (2014) describe the gamification of health through marketing apps such as Nike+Fuel Band. By quantifying everyday fitness with the tracking of movement, calories consumed/burned, and so on and then providing "rewards" for reaching specific thresholds, Nike's app has gamified physical fitness. As this example demonstrates, gamification has been criticized (particularly by educators) for overemphasizing extrinsic, temporary rewards such as achieving a new level or higher score (Knapp et al., 2014). The focus on extrinsic rewards, rather than intrinsically motivated open-ended play, tacitly embraces the hegemonic values of competition and hierarchy (Nolan and McBride, 2014). Further, gamification is increasingly used by companies offering a framework "that enables organizations to motivate employees and customers to do what they want done, by recognizing and rewarding their actions" (Fluss, 2015, p. 4). Software companies such as Bunchball have developed game applications, utilizing the competitive accumulation of points and leveling, to influence the behavior (e.g., productivity and brand awareness) of employees and consumers.

As a thoroughly digital experience, contemporary entertainment media, outside of video gaming itself, has been gamified in various ways. Most notably, sports spectatorship has been transformed from a largely passive form of entertainment into an interactive game via fantasy sports websites and other forms of digitally interfaced gambling. As a basketball fan, over the past few seasons, I found myself experiencing the NCAA men's basketball tournament in entirely new ways. The ESPN NCAA tournament mobile application encourages viewers to not only follow the results of their individualized "bracket" (picks of the winners and losers for every game in the tournament) but also the percentages of all users' correct and incorrect picks. The app sends frequent notifications, such as the following: "Of the over 13m brackets entered in the Tournament Challenge, just 14k (0.1 percent) brackets remain perfect after first day." The results are instantaneous and sent via text messaging, thus creating immediate numerical outcomes while watching tournament games. As I explain below, gamification is intimately tied to the design of objects that transform everyday life into a kind of ritualistic play.

Everyday Life and Everyday Things

Norman's (2013) influential work has examined the role of design in the pleasures and satisfaction associated with consumer products. At the simplest level, the popularity of the iOS is due to a well-documented achievement of design. Norman (2003) explains that "when done well, the results are brilliant, pleasurable products. When done badly, the products are unusable, leading to great frustration and irritation" (p. 5). I reference Norman's groundbreaking work to highlight the growing appreciation for design, particularly the idea that design "communicates" with users via a number of dimensions (e.g., affordances, signifiers, and mapping). These qualities of Apple's "hardware" are well documented in the field of design. Apple's reputation as a company almost obsessed with design is notable in its iconic simple white and black packaging. My intention is to go further inside the software of the iOS to elicit the experience of an individual user inhabiting everyday life. The iOS software is remarkable due to its simplistic and intuitive design. For instance, my mother, who in the past has refused to turn on her personal computer out of paralyzing fear it will break, can use her iPhone with great ease and confidence. Similarly, my children, even at the tender age of two, both could navigate the iPad with no instruction (sometimes grabbing it and playing when my wife and I were distracted) but struggle when trying to use our laptop computers. Of course, my experiences are far from unusual. It seems that for many users, iOS devices fit seamlessly into their everyday lives, an achievement that suggests as much about the software as the hardware.

As Norman (2013) observes: "Experience is critical . . . When our home technology behaves in an uninterpretable fashion we can become confused, frustrated, and even angry—all strong negative emotions. When there is understanding it can lead to a feeling of control, of mastery, and of satisfaction or even pride—all strong positive emotions" (p. 8). In particular, Norman describes a "visceral response" that can be frustrating and irritating, or in the case of the iOS, a sense of control and mastery within everyday life.

Influenced by the work of Lefebvre (2002) and De Certeau (1984), the routine habits of daily living, generally referred to as *everyday life*, have generated scholarship from varied academic disciplines (see Burkitt, 2004; Sandywell, 2004). Specifically, the time and space of everyday life represent essential issues concerning an individual's relationship to technology. The dimensions of everyday time and space suggest dialectical relationships swirling about the hypermediated subject such as an individual's experiences associated with *autonomy—institutionalized power*, *alienation—social interdependency*, *cultural order-transgression*, *extraordinary spectacle—routine banality*, *mass culture—media personalization*. As power is the fundamental force in these dialectical vacillations, everyday life is increasingly seen as a site of activism demonstrating the need to wed theory with practice (Pink, 2012). According to Pink, routine practices, such as the habits associated with domestic life, offer spaces for the reinforcement, codification, or resistance of oppressive power arrangements. Further, a common theme among these scholars is that self-reflexive participatory ethnographic practice is uniquely well suited to studying everyday life, especially the hypermediated experiences of daily living. In this project, I frame Apple iOS devices in terms of the dialectics of time and space within everyday life. Within this framework, the Apple iOS evoked Victor Turner's conceptualizations of everyday ritualistic practices and the ludic liminoid.

Everyday Life as Liminal/Liminoid Ritual

Before I can unpack Victor Turner's notion of the ludic liminoid, I must first examine the closely related and in many ways overarching term, liminality. Although an anthropologist, Turner, a bit of a kindred spirit with Kenneth Burke and Erving Goffman, often wrote about myth and symbolic action in a manner echoing the concerns of many communication theorists (for a particularly rhetorically slanted essay, see Turner, 1975). In a number of regards, the liminal fits seamlessly into Turner's larger project of exploring the routine practices of cultures as social dramas marked by collective performances and transitional changes (e.g., see Turner, 1980). Turner (1982) defined the liminal as an individual in a state that is "betwixt and between," or an experience of limbo. Turner (1982) explained:

> During the intervening phase of transition, called by van Gennep "margin" or "limen" (meaning 'threshold' in Latin), the ritual subjects pass through a period and area of ambiguity, a sort of social limbo which has few (though sometimes these are most crucial) of the attributes of either the preceding or subsequent profane social statuses or cultural states. (p. 24)

As a transitional and ambiguous experience, the liminal involves experimentation and discovery.

Liminal transitions and social dramas are closely tied to an essential cultural practice or form—the ludic (i.e., playful experimentation). As Turner (1982) emphasized, a unique cultural privilege (represented in the freedom of play) is offered by the marginality of liminality, which accounts for much of the pleasure and performative play of cultures: "to my mind it is the analysis of culture into factors and their free and 'ludic' recombination in any and every possible pattern, however weird, that is the essence of liminality, liminality *par excellence*" (p. 28). The playfulness of the ludic is central to the experience of the liminal as "liminality is peculiarly conducive to play" (Turner, 1980, p. 166). Further, according to Turner, playful cultural performances are inherently reflexive suggesting an interesting parallel with postmodern ethnographies of everyday life (described in Chapter 3), as participant observation fieldwork (particularly embracing the role of the flaneur) is sometimes described as an inherently liminal state (see Jackson, 1990). The concept of liminality has received some attention by researchers exploring communication processes such as the ritualistic practices of grieving (Jones, Zagacki, and Lewis, 2007) and performances of sexual identity (Lacroix and Westerfelhaus, 2005). As these examples demonstrate, liminality is conceptually well designed for fluid vacillation between identity-related binaries concerning gender, sexuality, and race as well as linear ritualistic rites of passage.

More specifically, digital culture has been described as liminal, particularly in terms of fantasy and identity play (Barbatsisi and Hansen, 1999; Denegri, Knott, and Molesworth, 2012). Shields (2000) observed that liminal rituals and the communication processes of digital environments share much in common:

> Whether by carnivalesque inversion of the social order, the liminal suspension of norms, or the queering of social regulation, the virtual is re-invoked and re-performed. Something of this relation can be found in both the history of the carnivalesque and liminality, as well as in the case of the virtual spaces of digital domains.

In the sense that *liminal* literally means "being-on-a-threshold" (Turner, 1979, p. 94), my experience with the iOS, like the experiences of other digital environments, is perpetually liminal or on the threshold or brink of *what is next*. More particularly, the iOS experience shares a kind of cultural family history with television remote control channel surfing or "grazing," a process of endless scrolling and anticipation (Bellamy and Walker, 1996). In order to be (in marketing language) "sticky" or encourage habitual use, the apps of the iOS often direct the user's attention to the constantly renewed menu of narrowly defined options (or in the language of games, a new level). For instance, my iPad apps offer endless anticipation of new content such as updated sports scores and news stories, the perpetually evolving news feed on Facebook, added levels to video games, or new music and movies in the iTunes store. As Turner (1979) states, liminal time "is a time of enchantment when anything might, even should, happen" (p. 94). In this regard, iOS devices "stage" liminal performances in public "quotidian spaces" (Turner, 1979, p. 96) by integrating a continuous sense of anticipation into the routines of everyday life. Nonetheless, as a placeless and ethereal postmodern experience, the everyday life practices of the iOS are not strictly liminal in Turner's modernist characterization of the concept. Therefore, I turn to the more postmodern concept of the liminoid.

The liminoid, as a narrower theoretical concept, differs from the liminal in several important ways. While still involving the transitional state of "betwixt and between," the liminoid is marked by individualized secular pursuits (rather than collective, sacred ritualistic practices) emphasizing commodities (e.g., lifestyle consumption). In this respect, liminoid phenomena "tend to be the leisure genres of art, sport, pastimes, games, etc." (Turner, 1982, p. 55). For instance, air travel is largely an experience of the "perpetual liminoid"—of fluid individualized performances immersed in a hyper-commodified environment with little sense of time or space (Murphy, 2002). Based upon Turner's (1979) description, it seems fair to suggest that liminality is principally a premodern and/or modern process, while the liminoid flourishes "in societies of more complex structure" (p. 116)—particularly uniquely postmodern cultures emphasizing individualized experiences immersed in contradictory discourses and dialectical forces. Like the liminal, the liminoid is "experimental in character," but liminoid phenomena are largely voluntary (unlike obligatory liminal rituals such as rites of passage) offering opportunities to "subvert quotidian and prestigious structures and symbols" (Turner, 1979, p. 116).

Unfortunately, the liminoid processes of postmodern everyday life have not been examined in necessary depth. Further, Turner's compelling concept of the liminoid remains relatively nebulous and abstract. Turner (1979) himself admitted that the liminal/liminoid conceptual distinction was "exploratory" and "crude" (p. 55). Based upon my auto-ethnographic observations, I believe the iOS experience is inherently liminoid in character as I am immersed in a transitional "on-the-threshold" state while navigating postmodern commercial media—such as anticipating the next *Avengers* sequel, analyzing the potential draft picks of an NFL team, and following the perpetually updated tweets of favorite celebrities. In this regard, in addition to explaining the immensely popular experience of the Apple iOS, this project further develops Turner's nascent concept of the liminoid.

IOS as Liminoid Play

Mobile devices such as the iPhone and iPad are associated with diverse experiences such as everyday talk to loved ones, planning meetings, and taking photographs. In this chapter, I am isolating one specific, overarching experience of the media logic of relevance: the liminoid (playful) flow-like experiences associated with clear rules and numerical outcomes. While other experiences are certainly a part of interacting with these devices, the majority of the popular apps and features of the iOS evoke consistent kinds of responses. Further, as I explore below, apps such as Instagram and Facebook transform everyday processes such as taking photographs and interacting with loved ones into a playful rule-based game with the quantifiable outcome of "likes." Similarly, the left/right swiping (to like or ignore potential romantic partners) of the popular dating app, Tinder, transforms romantic relationships, or at least the act of "hooking up" for a casual sexual encounter, into an interactive, game-like experience.

My embodied interaction with my iOS devices is fundamentally a playful, performative experience. As a starting point, my iPhone and iPad compel me to interact with the immersive icons ("apps") on its touch screen. In my interaction with the iOS, physical movement, particularly swiping the screen, touching apps, speaking (to others via the phone or to Siri, the voice command feature of the iOS), is integral to the experience. Further, the visual touch screen icons are chosen and physically arranged, probably based upon some deliberation, by an individual from the one million (and growing) apps in the Apple Store. At its most simple, the combination of touching and swiping the app icons is the essential design feature of the Apple iOS. In my observations, the intersection of app icons with physical movement was at the heart of the embodied iOS experience. Via movement within and across apps, I often find myself lost in the iOS universe, losing track of time and space—an experience remarkably similar to liminoid flow.

Toward the end of his career, working to develop the concept of the liminoid, Turner (1979) compared ludic liminoid experience to Csikszentmihalyi's influential notion of flow. Turner's discussion offers a compelling starting point for explaining the playful liminoid experience of the Apple iOS. Since its initial conception, flow has been applied in remarkably diverse ways—for this project, I am primarily interested in the phenomenological/experiential dimensions of flow that were at the center of Csikszentmihalyi's and Turner's work.

Csikszentmihalyi (1990) describes nine processes associated with flow: challenges and skills that are high and in balance, clear goals and feedback, concentration on a task, deep (but effortless) involvement, a sense of control, loss of self-consciousness, and a sense of alteration of the duration of time. Turner (1979) suggests that "liminoid metagenres" such as mountain climbing and chess can stimulate the experience of flow by directing and narrowing an individual's consciousness. In general, flow involves the "centering of attention on a limited stimulus field" (p. 56). As explored previously, the iOS can be best described as a frame (i.e., cognitive map)—the screen literally frames my stimulus field limiting and filtering my attention. Further, in a state of liminoid flow, the individual is completely in the moment, as consciousness is "narrowed, intensified, beamed" (p. 56). The intensive focus (demanding the individual's attention) often associated with iOS devices might be understood as a kind of flow-like experience. In general terms, Turner's (1979) definition of flow eerily echoes my embodied experience with the Apple iOS: "Flow is the holistic sensation present when we act with total involvement, a state in which action follows action according to an internal logic, with no apparent need for conscious intervention on our part" (p. 154). Flow is most effectively achieved when goals are clear and feedback is immediate (Shernoff and Csikszentmihalyi, 2009) which help produce "intense concentration and absorption in an activity with no psychic energy left over for distractions, a merging of awareness with action, a feeling of control, loss of self consciousness, and a contraction of the normal sense of time (i.e., time seems to fly)" (p. 132).

The iOS and the Gamification of Everyday Life

In my interaction with the iPhone and iPad, the iOS's promotion of flow-like experiences is tied to the gamification of mobile devices via apps. Video games have been strongly linked to the experience of flow in a wide range of research studies, particularly in the fields of psychology, information science, and education (Weibel et al., 2007; Shernoff and Csikszentmihalyi 2009; Cowley, Charles, Black, and Hickey, 2008). Specifically, video gaming is associated with interrelated experiences of flow, presence, and enjoyment (Weibel et al., 2007). Flow has received relatively little attention in communication and media studies (Jin, 2011), although initial experimental research suggests interesting connections between video game characteristics and flow (Jin, 2012). When reading Csikszentmihalyi's work, it is easy to recognize the features of flow in video games (e.g., concrete goals and manageable rules, clear numerical feedback, feeling of control, screening out distraction, and high/balanced challenges) (Sherry, 2004). As Sherry observed: "Video games possess ideal characteristics to create and maintain flow experiences in that the flow experience of video games is brought on when the skills of the player match the difficulty of the game" (p. 340).

For the purposes of this analysis, I focus on the experience of gaming and flow, rather than the myriad studies using flow to elicit specific outcomes (e.g., brand awareness for adver-gaming or curricular outcomes in schools). Specifically, in the field of media studies, flow has received some (albeit limited) attention. Most notably, scholars such as Weber, Tamborini, Westcott-Baker, and Kantor (2009) advocated for a "neurophysiological perspective" rooted in biological explanations of communication processes. Instead of physiological explanations, this project is interested in phenomenological and experiential understandings of flow and mobile devices. Flow offers an interesting means of answering a question vexing media scholars: Why do people enjoy media entertainment? (Sherry, 2004). Broadly speaking, media enjoyment of film, television, or novels involves "a balance between an individual's skill and the difficulty of the task at hand" (p. 336). Specifically, for video games, the appeal "results predominantly from the challenge of solving the puzzle presented in the game" (p. 338). Further, Csikszentmihalyi's work has not gone unnoticed in the multi-billion dollar video-gaming industry, as "the concept of flow has become a dominant paradigm in digital game design" (Stenros, Montola, and Waern, 2009, p. 155). Video game designers have known for decades that compulsiveness-associated video gaming is largely associated with calibrating the correct level of challenge and increasing the challenges with each new level. Of course, even the graphically simple video games of the 1980s such as Pac-Man and Donkey Kong began with relatively easy levels and steadily increased the challenge with each subsequent level, compelling players to plug quarter after quarter into the arcade games.

Liminoid flow is associated with three characteristics that are integral to the experience of many of my iOS applications: (1) a clear set of rules, (2) quantifiable outcomes and feedback, and (3) the feeling of control. First, in general terms, the iOS encourages flow-like experiences by establishing a clear set of rules within a limited perceptual field. When immersed in my iOS devices (iPhone, iPad, and iPod Touch), I have a limited set of choices, particularly when compared to other experiences of everyday life. For both Turner and Csikszentmihalyi, flow is most easily achieved with relatively rigid sets of rules such as in game playing. More specifically, in order to produce flow "within a computer-mediated environment, the goals must be clear for both the artefact (discussed below) and the task" (Finneran and Zhang, 2003, p. 488). The iOS (and all of its applications) of course is entirely a digital code or numerical representation (Manovich, 2001). Like traditional games such as chess with a clear set of well-defined rules, the ones and zeros of binary code provide the same rigid structure limiting an individual's attention—a particularly remarkable feat given the information overload characterizing postmodernity, as discussed in the previous chapter.

Steve Jobs described his design philosophy: "That's been one of my mantras—focus and simplicity. Simple can be harder than complex; you have to work hard to get your thinking clean to make it simple" (Valentino-Devries, J., 2011). Jobs' attention to simplicity, clarity, and focus pervades popular iOS applications. When I turn on my iPad, my choices are immediately framed by the apps on the screen. Within my favorite apps, I am further limited in my choices. Following the logic of digital code, apps typically offer a "this or that" (e.g., "like or dislike" on Facebook) series of options. For instance, the popular app for the Weather Channel immediately conjures local weather (based on location awareness) offering the user a simple choice between three links for "maps, news, or video." By comparison, the Weather Channel website (accessed on my laptop) offers a dizzying menu of choices with over forty individual hyperlinks. Apps for a wide range of websites such as news services (*New York Times* or CNN) or entertainment outlets (ESPN or Flixter) also tremendously simplify choices (when compared to their cluttered website versions) for iOS users. Overwhelmed by choices on websites, the rules imposed on Web surfers are relatively flexible, boundless, and fluid.

On the other hand, when using my iOS apps, much like with simple game play, the limitations offered by rules are relatively rigid and clear giving greater opportunities for flow-like experiences. Simple rules provide a greater sense of liminoid flow as an individual more easily "suspends disbelief" feeling "that rules are 'true'" (Turner, 1979, p. 58). For instance, on the Weather Channel website, there are a lot of video and hyperlinked stories about dangerous global weather (e.g., hurricanes and tornadoes) that invariably clutter the user's brain with worry and complicated questions. On the other hand, on the Weather Channel iOS app, the vast majority of this information is eliminated (filtered out) and the user *only* sees the current weather conditions and the five-day forecast for their city or town (and only has to choose between a few clearly displayed options).

Similarly, according to Turner, flow is also evoked via clear outcomes and feedback. In my experiences, many of the most popular iOS applications set up a number of both implicit and explicit numerical outcomes that consistently provide clear feedback. The most popular iOS games, which are among the most downloaded apps overall, such as Angry Birds, virtually always have numerical outcomes (sometimes called "ludus") ranking relative mastery (Frasca, 2003). In an astonishing coincidence, while developing this project, my daughter and her grandmother began intensively playing a popular iOS game titled *Flow*. The object of the immensely popular puzzle game is to connect colored dots with lines without breaking the path of any other colored lines. In the game, literally everything is quantified recording the level (based on the number of dots to connect), the number of moves, and previous top scores. Other popular apps have similar numerical outcomes such as the number of friends or followers on Facebook or Twitter and the number of songs in the iTunes library. In each case, a numerical outcome (number of songs, number of friends, number of likes, and so on) provides feedback to the user transforming, for instance, "web surfing" for information into a *game* of virally spreading news stories (with the winner spreading a story to the largest audience) via social networking. In other words, my iOS applications are filled with "subjective challenges" (Nakamura and Csikszentmihalyi, 2001) that help evoke a flow-like experience.

Gamification is clearly demonstrated by a popular traffic and navigation app, Waze. Tens of millions of people share their real-time traffic information (as gathered by the app via the smart phone), and Waze (which was purchased by Google in 2013) compiles this information to provide smart phone users the fastest driving route through their cities (Rodriguez, 2013). My wife and I started using Waze while visiting Las Vegas. We were attending a conference but staying at my stepsister's home (in a residential neighborhood) away from the conference hotel on the Las Vegas strip. This required us to navigate the congested Las Vegas freeways in our rental car. My stepsister suggested using

Waze during our commutes to the conference. Strikingly, the visual interface of the Waze app is like a video game. While following a blue icon along the route, bright cartoonish images representing accidents, construction, police, traffic congestion, and so on appear, with a bright purple line indicating an alternate route around the obstacles. Further, the precise arrival time appears prominently along the bottom of the screen representing a kind of score for the driver's efficient commute. The experience evokes a competitiveness seeking to find the fastest route around the obstacles (construction, traffic congestion, and so on) on one's daily commute. Further, as an app built on community reported information, individual Waze users can earn "points" for providing information in order to climb in rankings for the local community "scoreboard" (earning "levels" such as "Warrior Wazer"). Waze users are rewarded for the number of drives they mediate through the app, in the form of candy that appears on their screen (apparently consumed virtually as their car passes through it) or new avatars.

Liminoid flow is associated with a feeling of control (built upon subjective challenges) that compels worry and anxiety to disappear creating the perception of liberation. I believe the pleasures individuals feel concerning this perceived sense of power are striking relative to the previous experiences associated with everyday media objects. In the past, consumer electronic products involved largely passive experiences (perhaps evoking feelings of powerlessness). Television programs could only be viewed during rigidly scheduled times defined by a large corporation. Radio stations controlled playlists determining if and when songs were played. People needed to "wait by the phone" for an important call. As is so often highlighted in Apple's advertising, an individual user exclusively determines music playlists, movie playtimes and places, and texting/phoning patterns regardless of time or space. When I take a walk, listening to my selected playlists of songs and suddenly my wife calls, the slightest gesture allows me to pause a song, speak about our dinner plans, hang up, and seamlessly continue with my musical soundtrack. In this moment, I feel complete command over my environment. In my observations, sound powerfully evokes feelings of control. As Grossberg (1992) explained: "Music has a unique and striking relationship to the human body, surrounding, enfolding, and even invading it within its own rhythms and textures" (p. 152). The iOS is also a complex aural experience with sound effects and vibrations for calendar alerts, ring tones, and of course, my extensive iTunes music collection. The audio cues, combined with the physical movements of swiping and touching, elegantly remind me of my power to control when to reply to messages, what songs to play, which apps to initiate, and so on. Conversely, when the sound cues malfunctioned, such as when my music continued to play when I answered a call, I became immediately enraged while fumbling with my phone trying to stop the song, in hindsight, due to my feelings of helplessness.

For Csikszentmihalyi (1990), flow is inextricably linked to overall psychological happiness and well-being. If indeed flow produces the feelings associated with happiness, it is unsurprising that people describe a visceral discomfort associated with losing or forgetting their iOS devices. In my everyday life, the iPhone is a constant companion, almost another appendage. I find myself unconsciously and habitually checking for texts, e-mails, sports scores, and so on. This is not necessarily instrumentally motivated behavior—I do not always have an explicit preconceived goal; rather I am drawn to the *feeling* of interacting with the iOS. I desire to return to my affective relationship with my iPhone and iPad. If indeed this machine produces a sensation resembling happiness, it stands to reason that its absence would create discomfort. For critics of media-driven pleasure, this may resemble "so called junk flow or faux flow—with video games, reality tv shows, and idle gossip as prime examples. These have some of the elements of flow (engagement and absorbsion), but they are not especially challenging and they certainly do not leave us feeling invigorated or satisfied" (Peterson, 2006, p. 67). In comparison to Csikszentmihalyi's (1988) conceptualization of flow, unlike the specialized knowledge and skill required of chess players or rock climbers, the iOS can produce flow-like experiences for the relative novice as well as the highly skilled. For this reason, I deliberately use the term "flow-like" to describe my experiences with the iOS, as it has some important differences with traditional understandings of flow. Regardless of the authenticity of the flow-like experiences of the iOS, the iPhone and iPad's gamification of everyday, and the pleasures it produces, hold significant implications for media scholarship and an individual's everyday experiences.

CONCLUSIONS

As objects hundreds of millions of people use every day while moving about their everyday life, the iPhone and iPad have ushered in unique experiences, particularly related to gamification. I identified two central sources of pleasure associated my flow-like experiences with iOS devices with a wide range of implications: the aesthetics of technological design and ritualized transcendence via media objects.

The clear rules and numerical outcomes of gamified iOS apps perfectly realize the media logic of relevance by simplifying the complexities of a media-saturated culture. Within the framework of the present analysis, the gamification of everyday life is also the flowification of everyday life. iOS devices are everyday devices—intimately integrated into the time and space of my daily life. When routine, everyday tasks are mediated by game-like design, everyday life becomes a kind of simple mathematical puzzle to be solved. My mundane activities such as scheduling meetings on a calendar, maintaining social relationships, and listening to music involve game-like challenges encouraging flow-like experiences. In fact, HabitRPG, Inc has developed an iOS app titled "Habitica" that is designed to "gamify your life" by motivating "healthy," "hard-working," and "productive" everyday tasks such as eating a healthy diet, exercising, and efficiently completing work projects (i.e., a person's daily "to-do list") via numerical scores, rewards, community leaderboards, leveling, and the other common features of popular video games (Habitica Gamifies, 2015). When using these gamified apps, the iOS serves as both a sense-making cognitive map (chapter 4) for individuals overwhelmed by hypermediated everyday life and simultaneously a game-like experience with clear numerical feedback. In this respect, iOS applications filter, focus, and frame my consciousness within the narrow parameters of simple rules and quantifiable outcomes.

Not coincidentally, many of the features lauded by Norman (2013) as good design, such as feedback and the sense of control associated with clear rules, are echoed in the characteristics of flow. In this regard, iOS applications appear capable of transforming everyday life from drudgery (or, for many postindustrial professionals, unmanageable amounts of work) into the pleasures of play and flow-like experience. Due to complex economic changes, a growing number of people make their living not via manual labor but via the management and creation of information or intellectual capital (Teece, 2000). Generally, intellectual capital is everything an organization's employees "know" that has value to the organization (Taylor, 2001). Of course these economic changes are well documented and studied (an extensive survey is outside the scope of this project). For the purposes of this analysis, workers producing intellectual capital spend inordinate amounts of time staring at digital screens (personal computers and smart phones in particular) that serve to produce, record, store, and disseminate their intellectual capital.

Routine activities associated with intellectual capital production such as checking e-mail, recording activities in a calendar, networking with colleagues, and generally monitoring information such as stock prices and newspaper reports are filtered through the iOS. In this regard, iOS devices, promoting flow-like experiences, transform machines previously associated with the laborious long hours of intellectual capital production into a source of flow-like pleasure. Further, the iOS seamlessly integrates the public and private, allowing work to be intertwined with family and leisure life. In an interesting parallel trend, the productivity of American workers continues to increase while wages remain stagnant or even declining (Applebaum, 2012), generally with the willing consent of the American workforce. If in fact the gamified mobile applications of the iOS perform the magic trick of transforming arduous postindustrial work into playful ritual, it is actually quite unsurprising (although certainly ironic considering it is an instrument of their subjugation) to witness thousands of knowledge workers standing in line at Apple stores waiting for their new iPhone or iPad the day it is released.

The unique experience of the iOS is due to its aesthetic simplicity, which differs dramatically from the design logic of most contemporary technology that apparently seeks to include endless new and improved (and often confusing) options for every automobile, television, and dishwasher—what Norman (2013) calls "featuritis." Via the aesthetics of uncomplicated and clear shapes, sounds, colors, and so on iOS devices evoke flow-like experiences. As Apple's devices vividly represent, the experience of contemporary media extends well beyond the content inside the machine into the affective dimensions of the machines physical qualities (shape, color, sound, and so on).

As the second source of pleasure, interacting with my iPhone and iPad can sometimes produce a kind of transcendent state by encouraging performative practices such as ritualistic music and movement. In my observations, the sort of hypnotic state produced from some iOS applications was made vividly clear while watching my (then) three-year-old daughter sit and play with a series of simple cooking games on my iPad (Maverick software's Ice Cream Sundaes, Breakfast, Cookies, and so on). As I briefly mentioned in the introduction of this chapter, she entered a state of consciousness I had not witnessed previously, even during her fantasy play, dancing, singing, coloring, or other activities. The repetitive behaviors of the games created a kind of sublime calm that was eerily familiar—a state of consciousness I had experienced myself and observed in other iOS users. The only description that captures the experience is *trance-like*. As a father, the episode was more than little disconcerting. Further, the iPad remains my daughter's most-favored play object, preferred over her crayons, dolls, books, and other toys.

Communication scholars have argued that performance opens the door to transcendence (Bacon, 1996). Rites of passage and rituals, for instance, usher in significant changes to an individual's cultural roles. In this regard, for Turner (1979), ritual is a "transformative performance" (p. 64). Ritual, particularly when married to the experience of flow, induces the state of transcendence. As Ashley (1990) describes, "flow, which in Turner's model characterizes the liminal state, offers an experience of transcendence, a return to the primal condition of being" (p. 57). Via liminal ritual, transcendence is a kind of sublime escape from the confines of cultural roles, norms, and so on. de Roos (2008) similarly describes the transcendental power of digital experiences:

> Free to enter, free to generate new symbolic worlds of entertainment, sports, games and diversions of all kinds. It means freedom to transcend social structural limitations, freedom to play with ideas, with fantasies, with words, with

> paint, with social relationships, freedom to play in any way possible. Leisure activities are all optional and part of an individual's freedom, of his growing self-mastery, even self-transcendence.

While elaborate digital environments such as Second Life or World of Warcraft clearly offer alternative realities for liminal transcendental rituals, I believe that Apple's iOS may offer a kind of transcendence more intimately integrated into an individual's material everyday life.

A number of iOS design features work in concert to evoke a kind of transcendental state. For instance, in the iOS, music provides a perpetual soundtrack. Interestingly, Apple chose to integrate the same iPod music software into *all* iOS devices including the iPhone and iPad. When interacting with the iOS apps, once music is played (either from one's iTunes library or Internet radio apps such as Pandora or Spotify), it continues to play "in the background" as the individual accesses other apps. In fact, with wireless networks or bluetooth connectivity, once music begins on one's iPhone, it can seamlessly continue through a car or home stereo via airplay (features I utilize often). Like millions of others, music was something I enjoyed before my iPod, but after the iPod, music has become a constant companion in my life. Others have argued that music reframed in new contexts can provide moments of liminoid transcendence via identification (Taylor, 2012). More specifically, in liminoid rituals, such as rave/club spaces, music helps create quasi-religious transcendent experiences (St. John, 2004). Music is always accompanying my time immersed in iOS devices evoking random fleeting memories and, when ironically juxtaposed with my material environment (in my mind), commenting on my postmodern everyday life. A Talking Heads song will make the placelessness of an airport appear ridiculous, an Avett Brothers song will draw the inauthenticity of a Starbucks into relief, and a Nirvana song will recall a deep affection for old college friends. In addition, the sensual immersiveness of headphones sequesters my mind in the world of the iOS.

The pleasures of the design aesthetics and the transcendent state associated with Apple's iOS devices suggest varied implications for contemporary everyday life. Initially, the characteristics identified in this chapter (clear set of rules, quantifiable outcomes and feedback, and feelings of control) might be observed in other contemporary contexts in various forms such as automobiles (with digital interfaces) or home entertainment systems (the Apple TV and Apple's Carplay both essentially reproduce the iOS in these contexts). As iOS devices are integral to millions of individuals' everyday lives, the ludic liminoid appears to be at the center of twenty-first-century daily life.

Unlike the countless lifestyle advertisements that deceptively claim that a pair of running shoes or a soft drink will make a consumer *happy*, the Apple iOS is a consumer product that actually can reproduce an experience resembling happiness (at least in the social psychological meaning of the term championed by Csikszentmihalyi). In this regard, shopping apps (including the very popular Amazon and Groupon) evoke flow-like experiences for consumers. Groupon uses iOS notifications (which appear much like text messages) to indicate deals (coupons) with limited availability and time sensitivity. Shopping becomes a competitive game (with rules and quantifiable feedback) of finding the best deal on haircuts, restaurants, or concert tickets.

Nonetheless, by any reasonable standard, the "happiness" (pleasure) offered by this type of flow is, at best, fleeting. As Shields (2000) observed in other digital environments: "In contemporary society, liminality has been stripped of transformative power to become a commodified experience" (p. 13). In its present dominant practices, iOS devices offer a sort of amoral transcendence—devoid of a discernible moral system, except perhaps, the pleasures of technological progress and consumption (I explore this further via Ellul's concept of technique in the final chapter). This might help explain the necessity to replace iOS devices with the latest model. As Virilio (1991) has insightfully observed, the dehumanization of the "newest" and the "fastest" has replaced the humanization of the moral and enduring. The lasting transformation of liminal ritual has been supplanted by fleeting liminoid pleasures—feeling perpetually on the threshold. From this vantage, iOS devices do not appear to provide an enduring satisfaction, rather, an insatiable drive for *what is next.* The prevailing experiences associated with the iOS, as an object of ritualistic quasi-transcendence, whets the appetite of an individual with a desire that is never satiated.

Nonetheless, I do not wish to ignore the emancipatory potential of iOS applications and comparable liminoid postmodern experiences. As presently practiced, the iOS's apps are dominated by the fleeting pleasures of consumption and disposable commercial media, but these devices might be reconfigured for a wide range of other practices. As I explore in greater depth in the final chapter, understanding the pleasures associated with the iOS will open new opportunities for app design that cultivate values such as community involvement, democratic decision making, and ethical mindfulness. Further, flaneur-like self-reflexivity encourages potentialities for reconsidering the dominant practices associated with contemporary mobile applications. Liminoid ritual might be integrated into political activism, dialogue across cultural differences, public service campaigns, community art projects, and a host of other activities that could enrich local communities and individuals' everyday lives. Now that researchers and app designers understand the power bubbling under the surface of Apple's devices, we might unleash this power to ritualistically foster the communication practices essential to the health and well-being of individuals within contemporary cultures.

The interface with iOS devices is a profoundly embodied experience performed by individuals in the time and space of everyday life. In the 1990s, co-opting images of political activists such as John Lennon, Martin Luther King, Jr., and Gandhi, Apple foresaw the changes in the individuals' relationship to media devices with their advertising slogan, "Think Different." As a playful immersive experience, Apple's design compels unique relationships between individuals and consumer objects. In many respects, these machines not only compel individuals to *think different*, but also, in the time and space of everyday life, to *live different*. Specifically, as I explore in Chapter 6, the emergent conventions of mobile applications offer exciting possibilities for reconsidering the practices associated with mobile computing.

Chapter 6

Mobile Applications as Equipment for Living in a Hypermediated Culture of Technique

In the first decade or two of the existence of cinema, broadcast television, and popular recorded songs, somewhat unwittingly, an often rag tag group of producers, writers, performers, and directors developed emergent generic patterns that influenced audience's expectations for decades. For instance, today, listeners take for granted that a pop song is about three or four minutes long, despite the fact that vinyl single or "45" records (which could only fit about three or four minutes of music per side) have been replaced by compact discs and digital downloads (which could be of any length). Similarly, despite the emergence of online streaming services such as Netflix and diverse cable networks, the prime-time schedule of broadcast television networks has followed the same basic format for decades. In the second decade of their existence, I believe emergent generic norms are presently being developed for mobile applications (apps) that will likely influence the expectations of mobile digital technology for decades.

By filtering relevant information, mobile devices provide a lifeline that sifts through, limits, and simplifies the complexities of a culture immersed in screened data. Particularly, simple, gamified mobile applications, with clear rules and numerical outcomes, focus, frame, and filter an overwhelming media-saturated culture. In this regard, mobile applications often reproduce the efficiency and standardization that are the hallmarks of *technique* (an idea first articulated by Jacque Ellul in the 1960s). Technique is an all-encompassing worldview, tied to the pervasive practices associated with technology, promoting an everyday life dominated by quantifiable productivity. Ellul's notion of technique helps us better appreciate the potentially harmful presence of technology in our everyday life and the potential to interact more ethically and humanely with it. In other words, I hope iOS devices can fully realize the original advertising slogan of the iPhone: "There's an app for that. That's the iPhone. Solving life's dilemma one app at a time."

In Chapter 1, I raised two questions associated with persistent unresolved issues concerning mobile technology:

1. Despite the widespread knowledge of the dangers of texting (and using one's phone in general) and driving, why do people continue to use their mobile phones while driving automobiles?
2. After years of warnings from privacy advocates concerning the loss of freedom associated with revealing private, personal information on social media, via search engines, and the like, why do Internet users continue to willingly provide personal private information to Facebook, Google, etc.?

After the analysis of Chapters 4 and 5, I feel a deeper examination of these cultural patterns is in order. Bull (2004) argued that the space of the automobile was transformed by the introduction of the car radio. For Bull, the automobile is a "mediated soundscape" (p. 360). The placement of radios and stereos affects "habitual everyday notions of what it might mean to inhabit certain spaces" such as automobiles (p. 360). When a media device is introduced into a space, such as a radio into the automobile, the experience of living in that space is transformed. An automobile is an increasingly hypermediated space. Digital billboards fill freeways and highways. Automakers consistently herald how their new models seamlessly integrate mobile applications, MP3 players, cell phones, GPS, and Internet connectivity into the driving experience via ergonomic dashboard/steering wheel design and unique software. Not only are these mobile technology features the norm for new automobiles, but drivers report satisfaction (and thus consumer demand) with "connected" cars, particularly the integration of mobile smartphone applications into automobiles (Nielsen, 2015). Today, driving involves a multilayered cocoon with drivers immersed in two machines (the car and mobile applications) that radically compress time and space. Together, digital media devices and automobiles evoke a kind of "spatial and temporal hallucination" resulting from a "cinematic energy" (Virilio, 2004, pp. 119–120). When the cinematic hallucination of the automobile is coupled with the intensive focus associated with game-like applications, mobile applications fully immerse the consciousness of the individual. Given this unique cocktail of design features, it is certainly less than surprising that individuals use their smartphones while driving. The designs of mobile devices and automobile interiors are part of larger structural conditions in which screens demand an individual's attention as a sense-making tool (by clarifying one's identity, one's location, and the meaning of images) within a hypermediated culture. In simpler terms, people look at their phones while driving because, while experiencing a trance-like hallucinatory state, the mobile screen clarifies a cluttered sensory landscape.

The fleeting pleasures and postmodern sensemaking offered by mobile apps also explain why individuals so effortlessly provide private information to giant technology companies such as Google and Facebook. For instance, downloading their iOS application allows companies to mine significantly more data (by integrating with the personal information within the phone) when compared to simply visiting the company's website (Harnett, 2014). As noted previously, mobile apps are increasingly gamified evoking a flow-like experience. Further, apps, by collecting personal information, offer a personalized experience filtering information based upon the previous search or "click" patterns of an individual. Amazon knows exactly the type of shoes I prefer (Keen hiking shoes), Netflix knows the type of TV shows I enjoy (edgy dramas similar to *Breaking Bad*), and the Weather Channel knows my physical location at any moment. Corporations realize that the pleasures of mobile apps supersede the concerns about the privacy of consumers. Further, apps are largely marketed to individuals based upon efficiency and convenience. In a culture overcomplicated by media saturation, the clarity, focus, and simplicity of "efficient" mobile applications are irresistible. If I use my Amazon app, my user name, password, and credit card information are saved allowing for "one-click" shopping. If I remain logged into Facebook, I can easily and seamlessly upload a personal photograph from my camera app.

As I described in Chapter 2, contemporary society is so saturated by digital media that "unplugging" from technology based on nostalgic desires to return to an idyllic past, while perhaps comforting, is utterly unrealistic. I am aligned with scholars such as Deuze and Lash who believe the central question for the contemporary culture is "How can we more humanely and ethically interact with digital technologies that are so intimately and irreversibly integrated into everyday life?"

In this concluding chapter, I reconsider the problem of technology most famously articulated by Ellul, within the media-saturated contexts of contemporary culture. Particularly, further extending the case study of Apple's iOS devices, I examine the (perhaps ironic) potential of mobile applications as equipment for living in order to regain freedom in a technological society dominated by the quantification and dehumanization of technique. Media ecology theorists such as Ellul clearly articulate *the* problem of the technological society—the logic or worldview of technique inherently limits human freedom (via pervasive organizational structures, relational patterns, deeply imbedded values, and so on). I am exploring a kind of solution, or at least a potential disruption, in technique, based on Kenneth Burke's concept of equipment for living. In order to solve the various problems associated with mobile technology, we must understand the forces promoting a media logic of technique (particularly in an individual's consciousness) in everyday life and work to find better equipment for living within it.

A MEDIA ECOLOGICAL APPROACH TO HUMAN AGENCY

Broadly, the media ecology project is interested in the complex relationship between technology and culture or "the ways in which the interaction between media and human beings give a culture its character and, one might say, help a culture to maintain symbolic balance" (Postman, 2000). Specifically, Ellul (1964) believes the technological society is primarily associated with "the problem of freedom" (p. xxix) and that the cultural practices associated with technique dramatically limit human agency. Like many media ecologists, especially those aligned with the work of Ellul, I believe the fundamental ethical challenge of contemporary hypermediated society concerns a necessary renewal of reciprocal, moral, and empathetic human relationships at the interpersonal, social, and cultural levels of experience. In this regard, I advocate for a distinctly phenomenological humanism in the wake of media saturation in contemporary culture.

Ellul (1964) defines technique, the primary source of dehumanization in contemporary society, by distinguishing it from "machines, technology, or this or that procedure for attaining an end" (p. xxv). Rather than affiliating technique with the device, Ellul considers it "the totality of methods rationally arrived at and having absolute efficiency. . . in every field of human activity" (p. xxv). Technique is highly rationalistic, predicated on standardization, and focused on efficient functionality. Ellul (1964) states that "technique is the translation into action of [humanity's] concern to master things by means of reason . . . make quantitative what is qualitative, make clear and precise the outlines of nature, take hold of chaos and put order into it" (p. 43). Further, technique extends well beyond technology itself, as "technique transforms everything it touches into a machine" (p. 4), including human relationships, nature, and frankly anything else in its path. Ellul's description of technique evokes Kenneth Burke's famous description of humans: rotten with perfection. Within an environment dominated by the mathematical precision of binary code and quantifiable outcomes, the pursuit of perfection pervades my consciousness when using my iOS devices. As noted in Chapters 4 and 5, when immersed in gamified apps directing my consciousness, rationalistic quantifiable efficiency becomes my taken-for-granted reality. While it can promote simplicity, focus, and clarity, the ability of binary code to represent either/or choices also threatens to flatten reality and uphold the illusion of control on the part of the subject.

Historically, technology was often framed as antithetical to humanism (e.g., as dehumanizing), such as in Heidegger's landmark work. Fueled by a delusional hubris that humans can/should control nature and existence, individuals slowly lose their essential humanity (i.e., they become rotten with perfection). In very broad terms, based on these assessments in the twentieth century, humans should steer cultures away from technology (particularly technologies reproducing technique) in order to regain human freedom. As I explore below, and as noted previously in Chapter 2, I believe twenty-first-century culture has passed the tipping point in which individuals can turn away from or "turn off" technology. In other words, the proverbial genie is out of the bottle, and the question now is not whether humans should adopt technology, but rather how individuals can more humanely interface with the technologies already intimately integrated into the time and space of everyday life. Further, as explored in the previous two chapters and as underscored with equipment for living, the iOS offers users great potential to navigate the postindustrial context typified by "information overload" and the postmodern context threatening meaninglessness and placelessness for subjects.

Finally, Ellul distinguishes between technology (specific tools), which can be used and applied in diverse and alternative ways, and technique—a kind of hegemonic logic derived from dominant technological practices (e.g., organizing human relationships and economic transactions). As I explore further below, I believe aesthetics and art, practices inherently humane and skeptical of efficiency, functionality, automatism, and bureaucracy, offer opportunities for alternative applications of mobile technology.

TECHNIQUE IN A HYPERMEDIATED SOCIETY

Returning to the themes described in Chapter 2, the everyday practices and institutional processes of the contemporary culture are dominated by the logic of digital screens. Observing the contemporary technology-saturated culture, Deuze (2011) insightfully argues: "Our life is lived *in*, rather than *with*, media" (p. 138). Digital screens are everywhere all the time—in our pockets, filling our homes, at our work, in our cars, at restaurants, at airports, in children's toys, and so on. In the United States, the so-called digital divide associated with race, ethnicity, and age has been significantly bridged due to the incredible adoption rates of smartphones across demographic groups (US Census, 2014). Most significantly, smartphone ownership has reached virtual complete saturation among young people, suggesting that mobile application use will be the norm well into the foreseeable future. Every institution of the contemporary culture such as education, family, health care, religious groups, and government is deeply immersed in digital technology, not only offering websites but also developing (even integrating into) apps. As Livingstone (2009) concludes, digital technologies now "mediate every dimension of society" (pp. 1–2).

As I've articulated throughout the book, the looking glass of ubiquitous digital screens provokes a good deal of anxiety, manifest (for instance) in today's discourse of addiction or in postmodern theory's conceptualization of placelessness. Equally prevalent with technology is the inseparability of it from human experience. In 1954, Heidegger observed: "Thus we shall never experience our relationship to the essence of technology so long as we merely conceive and push forward the technological, put up with it, or evade it. Everywhere we remain unfree and chained to technology, whether we passionately affirm or deny it" (p. 287). In the 1960s and 70s, Ellul (1980) also recognized the ubiquity of technology as inescapable: "Technology then forms both a continuous screen and a generalized mode of involvement. Technology is in itself not only a means, but a universe of means—in the original sense of Universum: both exclusive and total" (p. 35). Ellul uses the example of weather forecasting, which is more profoundly mediatized in the contemporary technological culture. As noted in Chapter 4, an individual, standing outside, is likely to touch the Weather Channel app on a smartphone rather than interpret (via their embodied senses) the temperature, wind, clouds, and so on within the physical environment. In other words, "technology is in fact the environment of man" (Ellul, 1980, p. 38). The universality of technology means there is no escape—every dimension of everyday life across the entirety of the globe is immersed in this complex technological system.

In the contemporary culture, I believe the anxiety and ambivalence associated with technology, especially mobile digital devices, stem from the patterns and characteristics associated with the vast majority of mobile applications. Many of the most popular apps emphasize efficiency and productivity echoing the central characteristics of Ellul's technique. Seemingly foreshadowing contemporary mobile applications, Ellul (1964) argues that "technique must reduce [humanity] to technical animal, the king of the slaves of technique" (p. 138). The categories that dominate Apple's App Store (productivity, social media, games, and entertainment) are largely purveyors of technique. Most obviously, productivity apps are for efficient day planning quantifying every aspect of the individual's daily life. Games are designed to quantify experience in order to produce a numerical outcome ranking the individual's performance in relationship to others. As described in Chapter 5, this "gamification" transforms diverse experiences into game-like systems with clear rules and numerical outcomes (Giannetto, Chao, and Fontana, 2013, p. 197). In fact, gamification (via quantifiable outcomes) extends into a wide range of apps transforming everyday experiences such as health (via fitness apps), sports spectatorship (via fantasy sports apps), social interaction (via social media apps), and learning (via educational gaming apps) into numerical code with precise quantifiable outcomes. Increasingly, the game-like qualities (clear rules and numerical outcomes) of social media emphasize the efficient marketing of brands via targeted advertising. While gamification can focus, frame, and filter a complex media landscape, it also expands technique across the culture.

The gamification of work is a clear example of technique in everyday life. Corporations are utilizing video game design to influence workers in a wide range of activities such as sales techniques, interpersonal skills, and even health and fitness (Manjoo, 2014). Manjoo described an almost perfect manifestation of technique:

> Envision that every one of your professional endeavors was meticulously tracked and measured in points, that there were levels to complete and you were given prizes for excellence. That every workplace action provided a tangible sensation of winning or losing as part of a system engineered to keep you addicted, thrilled to come back every morning. (p. B1)

While working on this project, I was startled to discover that my daughter's weekly first-grade math homework was to complete a "level" on the online mathematics video game called "Frontrow." After she answers basic math problems correctly, she earns "coins" and thus the opportunity to move higher on the class leaderboard, which ranks all of the students in her class based upon their scores in the game. Coins also earn her the ability to shop for new avatars or to purchase accessories or clothing for those avatars (e.g., a pirate hat for her pink pig).

Built upon the technique of automatism and efficiency, within the logic of these gamified apps, individuals have little agency—remarkably the apps are used in uniform and standardized ways, essentially "conditioning" behavior (Ellul, 1964, p. 395). Ellul (1964) believed humans were being transformed into quantifiable data due to technique: "there was no longer a being in motion, but a point; not a series of acts, but a curve, a trajectory in abstract space and time" (p. 330). Further, as technique, these apps are completely amoral, never opening up opportunities for moral judgments of the conspicuousness of consumption, the (un)ethicality of media content, or the ultimate meaninglessness of a life dominated by efficient time management. Nonetheless, I believe the pervasive practices associated with these apps are based on generic patterns and habitual behaviors within the daily lives of individuals (i.e., everyday life). Furthermore, rather than advocate that we try to excise the looking glass entirely (in the spirit of recent science fiction movies), I believe that we might diminish technique in favor of human potential (as I elaborate below, with the aid of Burke's idea of equipment for living).

As I noted previously, despite widespread warnings from privacy advocates, individuals continue to frequent social media sites and search engines that collect personal private information. When an individual's experience is transformed into numerical data (via social media, professional transactions, and the like), everyday life is transformed into efficient exchange of quantifiable data with perceived value. Instead of disconnecting from this information society, increasingly, individuals are encouraged to view their private lives as data with value (Simonite, 2014). As I search for information about political candidates, basketball statistics, or a new pair of dress shoes, I am increasingly aware of my atomized profile of information designed to predict my consumer behavior, political affiliations, and even possible criminal activity. As I receive e-mails and pop-up ads for local concerts, sports apparel, and political causes, I feel profoundly reduced to binary code stored in a secure database somewhere in Silicon Valley. Further, my family and friends, when viewed on my screen, are also reduced to numerical representation and thus open to my surveillance. The iOS features apps such as Find My Friends designed entirely to surveil the movements of loved ones. Within the logic of technique, our intimate relationships are purely information in the form of binary code to be efficiently collected and analyzed.

Even my leisure activities such as spectator sports have been reduced to numerical outcomes. Professional sports teams, such as my favorite National Basketball Association teams, consistently use advanced metrics to quantify every detail of their players' performances including obvious dimensions such as shooting percentages on specific locations of the court and less obvious details such as sleep patterns and heart rate. (Berger, 2015). Sports teams are integrating wearable technology for even more advanced data about the everyday lives of athletes. Personnel decisions are increasingly based on the numerical scores measuring every detail of the player's life. From the fans' perspective, enormously popular fantasy sport sites such as DraftKings and Fanduel similarly reduce the player's in-game performance (based upon the player's statistical output) to a single number for the purpose of gambling and competing with faceless strangers.

When my students use classroom software platforms (currently, at my university, it is Blackboard), the programs numerically measure every detail of their performance. Staring at a screen for an online class with a quantifiable outcome for every moment the student has been on the website (e.g., minutes/seconds reading a document, the precise length, time/date of every posting to the discussion board, comparisons to college-wide averages, and so on), I realize my most humane professional activity, teaching, is deeply immersed in technique as my students and their learning have been reduced to binary code.

In his now classic novel *The Hitchhiker's Guide to the Galaxy*, Douglas Adams (1979) insightfully and humorously foretold the future when he wrote that, according to a fictional supercomputer, after seven and half million years of reflection, the answer to the great question "of life, the universe, and everything" is 42. Along these lines, revealing the values of the operating system (or at least the creators of the operating system) among Siri's (apparently randomized) responses to the question, what is the meaning of life?

- "I don't know but I think there's an app for that."
- "All evidence to date suggests it's chocolate."
- "I find it odd that you would ask this of an inanimate object."
- "42."
- "I can't answer that now, but give me some time to write a very long play in which nothing happens."
- "That's easy . . . it's a question concerning the purpose and significance of life or existence in general."

Siri's answers suggest that the underlying purpose of existence is somewhere between absurd meaninglessness and efficient automated technique. While obviously tongue and cheek, the significance of Siri's responses is illuminated when considering her appeal to children who have always known her constant presence in their lives. As my children have earnestly expressed on numerous occasions when offering questions of childlike wonder (the proverbial, "why is the sky blue?" type questions), "Ask Siri, Dad, she knows everything."

As explored in Chapter 2, a media-saturated culture raises complex questions for individual's daily experiences as examined via the concept of everyday life. As Lefebvre (2003) argues, power is concealed the banal routines of everyday life, "a depth beneath its triviality . . ." (p. 37). De Certeau (1984) describes the potential "tactical resistance" of oppressed individuals in their everyday lives, particularly via their manipulation and reappropriation of mediated texts and practical experiences. From this initial theoretical work, scholars have examined everyday life as an intersection of shifting notions of *time* and *space* (see Burkitt, 2004; Grossberg, 1992). For instance, as I introduced in Chapter 2, Coyne (2010) suggests that everyday life is largely meaningful due to emergent patterns or "repetition." In short, scholars of everyday life express (much like media ecologists such as Ellul) a tremendous concern for the eroding freedom of individuals in contemporary culture—specifically the diminished agency of an individual moving through the time and space of mediated society. Ikea eerily evoked these concerns with their unwittingly creepy advertising slogan, "Where the everyday begins and ends."

As a kind of elixir to the ills associated with the problem of freedom in everyday life, Kenneth Burke describes *equipment for living* as a tool to reframe and redefine the repetition and routine of everyday time and space. Steve Jobs imagined this possibility when he stated, "You'll see more and more perfection of that—computer as servant. But the next thing is going to be computer as a guide or agent."

BURKE'S CONCEPT OF EQUIPMENT FOR LIVING

Burke (1973) believed that literary forms such as comedy and tragedy offer the means for individuals to make sense of the complexities of everyday life. As equipment for living, art/literature is "designed to organize and command the army of one's thoughts and images" (p. 298). Ott and Mack (2009) explain that "for Burke, various literary forms are themselves ways of naming recurring social situations, thereby providing readers with symbolic resources for addressing those situations in their real lives" (p. 267). Similarly, Heidegger views art as offering opportunities to regain a connection to reality and spirituality in a technological society. In effect, art labels or names the ethical or existential issues in one's everyday life implying "a command (what to expect, look for)" (Burke, 1973, p. 294) directing or framing an individual's attention in time and space. For instance, proverbs offer "recipes for wise living" (p. 295) and "strategies for dealing with situations" (p. 296). When a "new set of typical recurrent situations" arises, new literary forms designed for navigating these experiences are created (p. 300). In terms of the looking glass, I view the development of aesthetic mobile applications as potential equipment for living in a media-saturated society.

I believe smartphone applications should be understood as an artistic form that has the potential to provide equipment for living in the contemporary media-saturated culture. Burke (1973) explains: "Art forms like 'tragedy' or 'comedy' or 'satire' would be treated as equipments for living, that size up situations in various ways in keeping with correspondingly various attitudes" (p. 304). To paraphrase Burke, what is important is not what art means but what art does (active). Art offers "strategies for selecting enemies and allies, for socializing losses, for warding off evil eye, for purification, propitiation, and desanctification, consolation and vengeance, admonition and exhortation, implicit comments and instructions of one sort or another" (p. 304). As I explore further below, the tremendous interactivity of the iOS and many mobile apps offer new ways to engage the time and space of everyday life via art. Reimagining mobile applications as an aesthetic form (rather than merely as productivity or efficiency) opens possibilities for individuals to construct complex meanings in the time and space of everyday life—throwing into relief the forces offering liberation and limiting freedom.

Of course the terms "art" and "aesthetics" open quite a hornet's nest of ideological connotations. For the purposes of this book, art and aesthetics are contrasted with Ellul's concept of technique. For instance, on one end of the spectrum, films can be cinematic with narratives that open moral questions expanding the audiences' sense of humane empathy (art). On the other end of the spectrum, film can also be propaganda (or "public relations") serving the efficient mobilization of war or consumer products (technique). It is the difference between paint by numbers (or filling in the blanks of a painting app on an iPad) and creatively expressing one's complex human experience with a paintbrush and watercolors (whether screened or not). It is the difference between reciting memorized responses on a standardized test and expressing earnest arguments and personal narratives in a classroom discussion. In short, within the framework of this book, art and aesthetics shift individual consciousness away from technique toward the basic values of human agency, moral decision-making, empathetic reciprocity, and creativity.

Mobile Applications as Equipment for Living

As argued previously, popular mobile applications tend to simply reproduce technique. The so-called choices contemporary digital technology affords the individual are often vacuous and meaningless—which celebrity gossip tweet to read, which formulaic pop song to download, which virtually identical pair of running shoes to purchase, and so on. As part of his final statement in *The Technological Bluff*, Ellul (1990) states, "by the sole aptitude for astonishment, we profit from the existence of little cracks of freedom" (p. 412). I believe, for individuals imbedded in a media-saturated society, these "little cracks of freedom" can erupt via aestheticized mobile applications. Artistic mobile apps have the capacity to unleash the individual from technique, but in its present nascent state, apps tend to inhibit freedom by directing an individual in the time and space of everyday life. Unfortunately, it appears that part of the appeal of mobile apps (especially productivity apps) is that they release an individual from the "burden" of decision-making in everyday life. In this sense, a reframing of mobile applications requires disruptions in the established patterns associated with mobile computing that have developed over the past decade. An aesthetic that transcends technique requires spontaneity, a connectedness to nature, moral judgments, the reproduction of culture, self-reflexivity, mystery, creativity, and above all, an expansion of human agency and choice.

Aestheticized mobile apps that transcend technique via the characteristics described above are relatively rare and, more importantly, difficult for ordinary people to find. Sadly, presently, in the iTunes app store, "art" does not have a distinct category on the home page. Upon searching for artistic apps, the individual is directed toward drawing and painting apps, primarily designed for children (and often serving as marketing for Disney or Viacom media products such as *Frozen* or *Dora the Explorer*).

A handful of artists and activists share the concerns associated with technique and everyday life outlined above. These concerns have led to compelling initial attempts at developing new orientations to the time and space of everyday life via mobile applications. For instance, Re+Public's augmented reality-based iOS app "resurfaces urban murals by overlaying digital content on to the physical environment. Through Re+Public's camera vision, urban murals trigger digital archives of murals that used to exist, but have been painted over, in addition to animated and interactive 3D digital art" (Seiler, 2015). The app interacts with public spaces in cities such as New York, Miami, and St. Louis. Specifically, in downtown Austin, Texas, the Recess Arcade Wall Interactive Mural, when engaged with the Re+Public app via an iOS device, depicts images of stone-faced information age workers surrounded by mechanistic gears evoking complicated questions about labor and existence. Via augmented reality and animation, buildings are transformed into virtual murals disrupting the individual's "efficient" movement across time and space (e.g., efficiently commuting to work for greater productivity using their Waze app). This enlivens public spaces such as train stations, office buildings, or bus stops, which are transformed into something not based on efficiency but on creating meaningful understandings of everyday time and space.

Similarly, the "No Ad" app "uses augmented reality technology to replace ads with artwork in realtime through your mobile device" (No Ad, 2015). Not only does the app throw the shear ubiquity of advertising in public spaces into sharp relief, but the "replacement" artwork is largely intended to critique mass consumerism (e.g., images of chains on hollow skeletons evoking notions of consumption as slavery) or reorient the individuals to others in their surroundings via deeply empathetic portraits of ordinary people. For instance, when holding one's phone up to an advertisement in a subway station, instead of a photoshopped image of a supermodel drinking vodka, the individual sees (on their iPhone) a stark, but vivid, portrait of a typical city dweller (with every blemish and wrinkle), someone that could be standing or sitting nearby. The stranger's humanity is immediately evident. The time and space of everyday life is invariably disrupted as an individual is shaken to reconsider efficient time management and reflect upon deeper meanings of place, otherness, and power.

The spatial and temporal movements of individuals within urban environments further clarify the stark distinctions between technique (that emphasizes efficiency) and the aesthetics associated with equipment for living. For instance, the Apple Watch (a mobile device worn on the wrist) is apparently primarily designed for productivity and quantifiable fitness/health outcomes. At its launch in March 2015, "saving time" (a phrase repeated by Apple CEO, Tim Cook, multiple times) was the device's primary selling point—that is, a more efficient everyday life guided by technique. Measuring movement and temporality precisely, an individual is strongly encouraged to view time and space in terms of mathematical efficiency (recall the metaphor of technology as "waste of time" from Chapter 1). As Ellul (1990) argues, "the more time we save, the less time we have" (p. 258). Ellul calls into question the so-called time management "uses" of various computer systems. An individual productively saves time, only then needs to "fill time" by mindlessly scrolling through celebrity tweets on Twitter, shopping for unneeded clothing on Amazon, or playing simplistic games such as Angry Birds.

On the other hand, public art apps bring hidden meanings within urban spaces into greater focus, transforming an individual's understanding of everyday life. For instance, Digital Sanctuaries offers a mobile application that uses audio recordings of poetry and jazz music as an "urban soundwalk" around Pittsburgh, Pennsylvania, inviting "the public re-engage with the built environment of a cityscape" (Reese, 2014). Instead of viewing an urban space as merely designed for efficient productivity (commuting) or conspicuous consumption (shopping), the physical environment of the city is transformed opening new meanings for time and space tied to history, architecture, sound, and visual art. Similarly, the Denver poetry map invites local poets to write short poems about ordinary places (not simply historical landmarks) that can be read (via Google maps) as individuals move about the city. Such an app is particularly promising in humanizing the postmodern condition, interrupting efficiency with inspiration, and placing people in relation to one another in a multidimensional way.

As Ellul and Lash (among others) have argued, it is difficult to self-reflexively critique technology while immersed in technology, but we must try. Reconsidering the issues raised in Chapter 2, Lash (2002) argues that "we can no longer step outside of the global communications flow to find a solid fulcrum for critique. There is no more outside. The critique of information is in the information itself" (p. 220). Similarly, Ellul (1990) contends that "Technique is here, it is our environment. Can we abstract ourselves from it? That is our problem" (p. 132). A world of hyper-real images referencing other images in a sea of simulation (Baudrillard, 1994) demands artistic forms that rupture little cracks of freedom and perhaps expand them (or invite others to begin the cracking themselves). I believe apps based in aesthetics offer opportunities for reflection and critique bringing the hyperreal into discernible focus inviting new, self-reflexive meanings. The apps described above transform public environments often seen as meaningless, repetitive omnitopias (Wood, 2009) into places for meaningful existential self-reflection—particularly about the role of technology in an individual's life. These aesthetic apps suggest new ways to (re)orient to the time and space of everyday life transforming public environments into literary forms such as satire and tragedy. While commendable first attempts to disrupt technique, the apps outlined above are far from perfect realizations of equipment for living in a hypermediated culture. Drawing upon Harold's (2007) work, these apps could invite greater communal participation and collective creativity utilizing the full potential of *interactive* digital media. Regardless, I celebrate these artists' creativity and defiance of the conventional discourses that fatalistically lament the individual's meaningless "addiction" to technology.

As noted in Chapter 2, simply longing for an idyllic, nostalgic past devoid of mobile technology does little to address the dilemmas facing a culture that has fundamentally integrated these devices at both the institutional and personal levels. A number of years ago, major live performances such as music concerts and sporting events would discourage the use of mobile devices during the performance. Today, recognizing the unavoidable presence of these devices, artists and event venues now creatively integrate mobile phones into performances. Sporting events incorporate fans' selfies into the jumbotron images. Musicians encourage fans to use their phones in darkened arenas to create collective light shows. As these relatively simple examples demonstrate, via self-reflexivity and creativity, rather than simply stigmatizing these machines, we can harness the power of these ubiquitous devices for unique cultural experiences—ones that extend beyond minor enhancements of live events.

Aesthetic apps suggest the potential of mobile digital devices for expanding an individual's consciousness of time and space in hypermediated culture. The next step is to develop apps that further evoke empathetic and reciprocal human relationships/communities via aestheticized experience. I challenge software designers and artists to collaborate in order to create mobile applications that serve to evoke literary and artistic forms for moral sensemaking in the media-saturated environment of everyday life.

CONCLUSION

While it sometimes feels as though smartphones have been around for decades, in fact, the ubiquity of these machines is still very much in its infancy. After endless cajoling from a number of friends, recently, my wife and I finally got around to watching the HBO series, *The Wire*. It was striking that a show that ended only a few years ago had a conspicuous absence of smartphones. Similarly, as a teacher, whenever I play a scene from a movie, documentary, or television show from only ten or twenty years ago, contemporary students chuckle in the back of the room at the "flip" or "brick" phones used by the characters. The point is, humans are still creating the norms and genres for these nascent machines and deciding how to best interact with these devices. In this regard, much like the early years of cinema and recorded music, I believe we are at an exciting crossroads in terms of the patterned expectations of mobile digital devices. As I set forth in the introduction, the common sense of treating our immersion in the looking glass as an addiction or as a harmful distraction from "real" (immediate or face-to-face) communication must be replaced.

Yet, we must be wary of tech companies and app developers who may replace our common sense view of screens with technique. In recent years, the press has paid substantial attention to mobile apps that limit "anti-social" or self destructive behavior such as substance abuse, obesity, juvenile delinquency, and the like. Further, software designers have even developed "self-help" or "wellness" apps for parenting, relationship maintenance, and health/fitness (managing hypertension, diabetes, pregnancy or other chronic health-related conditions). Unfortunately, these apps uniformly transform human experience into quantifiable outcomes and streamlined efficiency, quite clearly reproducing technique in the everyday lives of the smartphone users.

Based upon the pervasive ambivalence and dread associated with technology, I believe individuals are increasingly aware that digital screens significantly frame everyday experience. This consciousness of the perceptual-focusing function of mobile devices coupled with an understanding of how apps evoke fleeting pleasures will hopefully direct individuals to more carefully choose the apps that fill their phones and other mobile devices. Further, individuals can encourage the design of apps that better serve the quality of their everyday lives by providing equipment for living, rather than reproducing apps that further extend the insidious reach of technique.

Despite how technique presently dominates mobile applications, I can imagine tremendous aesthetic possibilities for mobile apps that are intimately connected to the time and space of everyday life. In addition to the mobile applications highlighted in this chapter, a wide range of unique apps are potentially capable of disrupting the technique dominating contemporary culture opening new opportunities for moral everyday practices. For instance, while flow is often conceptualized as an individual or isolating experience, with particular types of software design, flow can be experienced socially and cooperatively (Raphael, Bachen, and Hernández-Ramos, 2012). Similarly, innovative forms of video game design (and more broadly gamification) have fostered community-driven creativity and problem-solving (Green and Kaufman, 2015). Finally, researchers have developed smartphone applications to record individual's moral decision-making in routine daily life (Miller, 2014). Together, these examples suggest that mobile applications can be developed that profoundly defy the dominant practices associated with mobile devices. Mobile applications can move beyond the simplistic individuated gratification of moving up levels of Candy Crush (and other popular apps such as Waze or Groupon that follow this same basic design) into communal practices that create artistic expressions evoking self-reflexive understandings of everyday life.

In order to disrupt technique, when digital technology evokes flow-like experiences and functions as a cognitive map, it should extend beyond individuated satisfaction and instead promote the values of community, empathy, reciprocity, and so on. If individual satisfaction is the end in and of itself (rather than a means of building a more ethical culture), one can imagine a dystopian everyday life of people sitting alone in cubicles feeling the fleeting pleasures of moving up levels of gamified environments for their work, social life, romantic encounters, education, and political discourse. Instead of fleeting, selfish pleasures, mobile applications have the potential to promote moral, creative, and cooperative decision-making within one's community. Taking this a step further, mobile applications appear capable of tackling many of the greatest problems of the twenty-first century by stimulating democratic discourse, empathy, and dialogue across cultural differences, innovative forms of problem solving, and commitments to diverse local communities.

In 2014, two popular films, in quite different ways, disrupted hegemonic orientations to everyday life. *Boyhood* hauntingly portrayed the everyday memories of childhood compelling the viewer to reconsider time, space, and aging in (at least for myself) deeply profound ways. Concurrently, *Birdman* fused imagery of disposable Hollywood blockbuster films into an aging actor's existential crisis raising questions about the centrality of celebrity and mass entertainment in the viewer's everyday life. These two films, both financially and critically successful, suggest that, in a cinematic era dominated by superhero franchises, sequels to prequels, and CGI blockbusters, unconventional artful movies can still be produced and reach a mass audience. For myself, both films served as equipment for living within a complicated postmodern society providing meaning to daily experiences (e.g., the tragic aging of my children and the satirical absurdity of my popular culture consumption). Similarly, despite the pervasive productive and efficient practices associated with mobile devices, I believe artful mobile applications can also emerge that function as equipment for living and transcend the hegemony of technique. At the end of *Boyhood*, the main character, Mason, who the audience has literally witnessed aging over his entire childhood, reflects: "Yeah. I know, it's constant, the moment, it's just, it's like it's always right now, you know?" In *Birdman*, at the height of the lead character's personal and professional (as an actor and director) crisis, an erratic man on the street quotes Shakespeare's prophetic words: "Life's but a walking shadow, a poor player that struts and frets his hour upon the stage and then is heard no more. It is a tale told by an idiot, full of sound and fury, signifying nothing."

Technology cannot be fully excised from our lives. Unbeknownst to my wife and I, our three-year-old son discovered Siri, the famous voice response program for the iPhone. I had left my phone on the kitchen table and he, being fascinated by our phones, pressed the prominent home button. Having seen my wife and I talk to Siri many times before, he said, "Siri, play Vampire Weekend." The phone began to play music from the indie rock band he enjoys listening to while riding in the car. This began my son's love affair with Siri. He adores talking to Siri and thinks of her as a real person who says funny things after he presses the iPhone's home button and asks, "Siri, tell me a joke!" The smart phone is a remarkably familiar object to a contemporary preschooler. In addition to *Dora the Explorer* (described in Chapter 2), PBS shows such as *Peg + Cat* (a family favorite) and *Superwhy* feature young characters interacting with smart phones on a consistent basis. Like many other kids, as a toddler, my son received no less than three toy "smart phones" (designed to look like oversized plastic iPhones) from members of our family. At first glance, the logic seemed sound—knowing that the toddler will be fascinated by a dominant object of their mom and dad's attention, simply offer the toy phone as a substitute. Of course, as most parents discover quickly, the child (even a toddler) clearly knows the difference between the toy and the *real* iPhone and rapidly loses interest in the toy.

As I am writing the conclusion to this book, it is apparent that technology in general and mobile technology in particular will change radically over the next few decades. Presently, significant commercial applications for virtual reality, artificial intelligence, and wearable technology are emerging at an accelerated rate. As these new technologies enter people's daily experiences, I hope we can self reflexively adopt these devices by thoughtfully considering how they will impact the fundamental practices of everyday life. While waiting for artists and software designers to unleash the full aesthetic potential of mobile applications, individuals can embrace the self-reflexivity of the flaneur to make sense of their experience in the media-saturated culture. In practice, this requires an individual to consciously disrupt the mindless routines that so effortlessly flow (pun intended) from mobile digital devices. Although a personal process, I encourage readers to model practices outlined in this book by journaling their reflections about digital technology, juxtaposing (ironically) in collage form the contradictory images floating about their devices and allowing emergent, fleeting moments to form meaningful narratives. Through mindful practice, individuals can more deeply understand the unique role that technology plays in their lives.

In 2013, comedian Louie CK appeared on Conan O'Brien's talk show. He told a long story of sitting in his car, listening to a Bruce Springsteen song, feeling the emptiness and despair of his mortality. Instead of reaching for his phone to relieve the overwhelming emptiness, Louie described choosing to instead embrace his lonely angst. He argued that smartphones prevent people from feeling deeper emotions and thinking profound existential thoughts because, when using a mobile device, "You never feel completely sad or completely happy. You just feel kind of satisfied with your product. And then you die." Returning to the central focus of this book described in Chapter 1, in the present moment, people feel an intense bittersweet ambivalence toward mobile technology. We love the moments of fleeting pleasure but deep down realize something is missing.

Instead of a portal (as in Alice's adventures in wonderland), perhaps the proverbial looking glass of our mobile devices should be seen as its literal meaning, a mirror. Staring in the mirror is often a means of self-examination (an image pervasive in pop songs such as Michael Jackson's "Man in the Mirror" or Elvis Costello's "Deep Dark Truthful Mirror"). Presently, individuals cling to their mobile devices as a cognitive map for making sense of a culture dominated by digital screens. Further, the devices evoke flow-like pleasures for brief moments of gratification largely inducing the efficiency of technique. In the case of mobile devices, the looking glass might be expanded into a kind of cultural mirror. Many of us are a lot like my preschool-aged old son, mindlessly pressing buttons on mobile devices anticipating a fleeting instance of pleasurable gratification. By embracing the self-reflexivity of the flaneur, especially when evoked by applications drawing attention to the time and space of everyday life, individuals will recognize the ethical dimensions of their life immersed in screens.

CODA

Every night for the past seven years, I have sat in a completely darkened and silent room, what is essentially a sensory deprivation chamber, for thirty to sixty minutes (or more). This was not by choice. I should explain. Talking with other parents, every child seems to struggle with at least one thing. Compared to other families, our kids have had a relatively easy time with toileting, breastfeeding, eating solid food, walking, talking, and reading, but both have struggled mightily with sleep. Since they were infants, our kids have been restless and resistant sleepers. After trying almost every sleep training method, we found that (after a sufficient amount of reading of picture books) if I simply sat next to them at night (and during naps), my calming presence would relax them into a deep sleep. We were thrilled to find a solution after many, many sleepless nights. The unintended consequence of our sleep solution is a daily dose of solitude in a culture of cluttered hypermediation. During this bedtime routine, I missed (immediately) receiving countless texts and e-mails as well as numerous live television broadcasts including the end of seven Super Bowls, countless basketball games, the series finales of *Breaking Bad* and *Mad Men* (among others). In fact, as a resident of the city of Denver, when the Broncos won the Super Bowl last year, I experienced their victory while sitting in darkened silence next to my three-year-old son hearing the cheers, fireworks, and honking horns outside on the street. Even in a dark and quiet bedroom, digital media was still persistently dominating my consciousness—I typically replied to the missed texts and e-mails and watched these television programs (via DVR) later that same night.

Honestly, the vast majority of nights resting with my kids, stuck in a mindset of efficient productivity, I simply mentally catalogued the tasks I needed to complete, usually racing to my iPad or computer after they dozed off. But, once in a while, on a few occasions, when I allowed my mind to disrupt the immersive grip of technique, I was able to deeply reflect upon my life and, particularly, the role of mobile technology. Several significant ideas in this book came in these brief, sporadic moments of solitude next to my sleeping children. Perhaps, this is one of the greatest sources of anxiety and angst chronicled in Chapters 1 and 2—a loss of quiet solitude or as Louis CK described sitting alone without his phone to Conan O'Brien, "that's being a person, right?" Before mobile technologies, every person had many moments throughout their day alone in their thoughts—walking or driving to work, waiting for a train or bus, doing simple tasks such as washing the dishes.

Once inside the universe of our own thoughts, at least in my experience, the priorities of life are all too obvious—our relationship to the people we love, our health and well-being, the safety and vitality of our communities, and so on. I am fully aware that as my children grow older and more independent, my moments of compulsory evening solitude will disappear and I will again be perpetually immersed in digital media. Further, short periods of sensory deprivation do not viably or effectively disrupt the ever-present and persistent consciousness of technique. As I conclude this book and move forward in my life, I am committed to finding ways to creatively engage mobile technologies that enrich my life and the lives of those around me. I hope this book serves as a call to every reader to reimagine the role of mobile technology in everyday life discovering yet unknown applications that renew the ethical, humane, and communal dimensions of our lives.

Bibliography

Adams, D. (1979). *The hitchhiker's guide to the galaxy*. New York: Harmony Books.

Allan, K. (2011). *Contemporary social and sociological theory: Visualizing social worlds*. Thousand Oaks, CA: Sage.

Anderson, M., & Caumont, A. (2014, September 24). *How social media is reshaping the news*. Pew Research Center. Retrieved from http://www.pewresearch.org/fact-tank/2014/09/24/how-social-media-is-reshaping-news/

Anderson, M. (2015, October 29). Technology device ownership. Pew Research Center. Retrieved from http://www.pewinternet.org/2015/10/29/technology-device-ownership-2015/

Apatow, J., Ferrell, W., McKay, A. DreamWorks Home Entertainment (Firm). (2004). *Anchorman: The legend of Ron Burgundy*. United States: DreamWorks Home Entertainment.

Applebaum, E. (2012, August 25). Report gives new insight in decline of the middle class. *US News & World Report*. Retrieved from http://www.usnews.com/

Ashley, K. (1990). *Victor Turner and the construction of cultural criticism*. Bloomington, IN: Indiana University Press.

Atler, C. (2014, January 9). Prime time out: 44% of Americans use gadgets while watching TV. *Time*. Retrieved from http://entertainment.time.com/

Auge, M. (1995). *Non-places: Introduction to an anthropology of supermodernity*. London: Verso.

Auslander, P. (1999). *Liveness: Performance in a mediatized culture*. London: Routledge.

Bacon, W. (1996). The dangerous shores: One last time. *Text and Performance Quarterly*, 16, 356–358.

Barbatsisi, G., & Hansen, K. (1999). The Performance of cyberspace: An exploration into computer-mediated reality. *Journal of Computer-Mediated Communication, 15*. Retrieved from http://jcmc.indiana.edu/vol5/issue1/barbatsis.html

Barthes, R. (1964). *Elements of semiology*. New York: Hill and Wang.

———. (1972). *Mythologies.* Translated by Annette Lavers. New York: Hill and Wang.

———. (1977). *Image-Music-Text.* Translated by Stephen Heath. New York: Hill and Wang.

Bauder, D. (2006, September 21). Average home has more TVs than people. *Washington Post.* Retrieved from http://www.washingtonpost.com/

Baudrillard, J. (1988). Simulacra and simulations. In. M. Poster (Ed.), *Jean Baudrillard: Selected writings* (pp. 166–184). Stanford, CA: Stanford University Press.

———. (1994). *Simulacra and simulation.* Ann Arbor: University of Michigan Press.

Beahm, G. (2013). *I, Steve: Steve Jobs in his own words.* New York: Collins Business.

Bellamy, R., & Walker, J. (1996). *Television and the remote control. Grazing on a vast wasteland.* New York: Guilford Press.

Berger, K. (2015, June 3). Warriors wearable weapon? Devices to monitor players while on court. *CBS Sports.* Retrieved from http://www.cbssports.com/

Benjamin, W. (1968). *Illuminations: Essays and reflections.* New York: Schocken Books.

———. (1999). *The arcades project.* Cambridge, MA: Harvard University Press.

Berryman, M. (2011). This changes everything. *The Design Management Review, 22*, 62–68. doi: 10.1111/j.1948-7169.2011.00158.x

Blair, C., Jeppeson, M., & Pucci, E. (1991). Public memorializing in postmodernity: The Vietnam Veterans Memorial as prototype. *Quarterly Journal of Speech, 77*, 263–288. doi:10.1080/00335639109383960

Blair, E. (2009, December 31). In this decade, every room is a screening room. *National Public Radio.* Retrieved from http://www.npr.org/

Bochner, A., & Ellis, C. (1992). Personal narrative as a social approach to interpersonal communication. *Communication Theory, 2*, 65–72. doi: 10.1111/j.1468-2885.1992.tb00036.x

Borchard, K. (1998). Between a hard rock and postmodernism: Opening the Hard Rock Hotel and Casino. *Journal of Contemporary Ethnography, 27*, 242–239. doi: 10.1177/089124198027002004

Buck-Morss, S. (1977). *The origin of negative dialectics.* New York: The Free Press.

———. (1989). *The dialectics of seeing: Walter Benjamin and the arcades project.* Cambridge, MA: The MIT Press.

Bull, M. (2000). *Sounding out the city: Personal stereos and the management of everyday life.* Oxford: Berg.

———. (2004). Automobility and the power of sound. *Theory, Culture, & Society, 21*, 243–259. DOI: 10.1177/0263276404046069

———. (2007). *Sound moves: Ipod culture and urban experience.* London: Routledge.

Burke, K. (1969). *A grammar of motives.* Berkeley, CA: University of California Press.

———. (1973). *Philosophy of literary form.* Berkeley, CA: University of California Press.

Burkitt, I. (2004). The time and space of everyday life. *Cultural Studies, 18*, 211–227. doi: 10.1080/0950238042000201491

Carbaugh, D. (1996). *Situating selves: The communication of social identities in American scenes*. Albany, NY: State University of New York Press.

Carbaugh, D., & Hastings, S. (1992). A role of communication theory in ethnography and cultural analysis. *Communication Theory, 2*, 156–164. doi: 10.1111/j.1468-2885.1992.tb00035.x

Carey, J. (1989). *Communication as culture: Essays on media and society*. Boston: Unwin Hyman.

Carroll, L. (2015, July 29). Yes, texting while walking is actually dangerous say study, and common sense. *The Today Show*. Retrieved from http://www.today.com/

———. (1946). *Alice's adventures in Wonderland ; and, through the looking-glass*. New York: Grosset & Dunlap, Inc.

Caron, A. & Caronia, L. (2007). *Moving cultures: Mobile communication in everyday life* . Montreal: McGill-Queen's University Press.

Carrington, V. (2012).'There's no going back'. Roxie's iPhone®: An object ethnography. *Language and Literacy, 14*, 27–40.

Castells, M., Fernandez,-Ardevol, M., Qiu, J., & Sey, A. (2007). *Mobile communication and society: A global perspective*. Cambridge, MA: MIT Press.

Chang, A. (2012, May 15). It's official: Cellphone owners love Apple hardware. *Wired*. Retrieved from http://www.wired.com/

Connor, C. (2015, July 31). Wasting time at work, the epidemic continues. *Forbes*. Retrieved from http://www.forbes.com/

Conover, M.D., Ratkiewicz, J., Francisco, M., Goncalves, B., Menczer, F., & Flammini, A. (2011). Political polarization on Twitter. Fifth International AAAI Conference on Weblogs and Social Media.

Conquergood, D. (1991). Rethinking ethnography: Towards a critical cultural politics. *Communication Monographs, 58*, 179–194. doi:10.1080/03637759109376222

Couldry, N. (2008). Mediatization or mediation? Alternative understandings of the emergent space of digital storytelling. *New Media & Society, 10*, 373–391. doi: 10.1177/1461444808089414

Cowgill, C. (2015, March 6). Special series: Unplugging from technology. *PBS Media Shift*. Available: http://www.pbs.org/

Cowley, B., Charles, D., Black, M., & Hickey, R. (2008). Toward an understanding of flow in video games. *Computers in Entertainment, 6*, 1–27. doi: 10.1145/1371216.1371223

Coyne, R. (2010). *The tuning of place: Sociable spaces and pervasive digital media*. Cambridge, MA: MIT Press.

Craig, R. (1999). Communication theory as a field. *Communication Theory, 9*, 119–161.

Crawford, L. (1996). Personal ethnography. *Communication Monographs, 63*, 158–170. doi: 10.1080/03637759609376384

Cronin, B. (2015, June 11). At 150, Alice in Wonderland is still popular. *The Wall Street Journal*. Retrieved from http://www.wsj.com/

Csikszentmihalyi, M. (1988). The flow experience and human psychology. In M. Csikszentmihalyi, & I.S. Csikszentmihalyi (Eds.), *Optimal experience:*

psychological studies of flow in consciousness (pp. 15–35). New York: Cambridge University Press.

———. (1990). *Flow: The psychology of optimal experience*. New York: HarperCollins.

———. (2000). Flow. In A. Kazdin (Ed.), *Encyclopedia of Psychology, Vol. 3* (pp. 381–382). New York: Oxford University Press.

Cvoro, U. (2008). Dialectical image today. *Continuum: Journal of Media and Cultural Studies*, *22*, 89–98.

Darlin, D. (2005, November 19). How to tame an inflated entertainment budget, *New York Times*, Retrieved from http://www.nytimes.com/

de Certeau, M. (1984). *The practice of everyday life*. Berkeley, CA: University of California Press. (Steven Rendall, Trans.)

de Roos, H. (2008). The transforming potential of second life. Retrieved from http://www.religieweb.nl/scriptie/second-life.html

DeLuca, K., & Peeples, K. (2002). From public sphere to public screen: Democracy, activism, and the "violence" of Seattle. *Critical Studies in Media Communication, 19*, 125–151. doi: 10.1080/07393180216559

Denegri-Knott, J., & Molesworth, M. (2012). *Concepts and practices of digital virtual consumption*. New York: Routledge.

Denver Poetry Map. (March 4, 2015). Mapping the city through poetry. Available: http://www.denverpoetrymap.net/

Denzin, N. (1992). *Symbolic interactionism and cultural studies: The politics of interpretation*. Cambridge: Blackwell.

———. (2006). Pedagogy, performance, and autoethnography. *Text and Performance Quarterly, 26*, 333–338. doi: 10.1080/10462930600828774

Debord, G. (1994). *The society of the spectacle*. New York: Zone Books.

Deterding, S., Dixon, R., Khaled, R., & Lennart, N. (2011). From Game Design Elements to Gamefulness: Defining "Gamification". *Proceedings of the 15th international Academic MindTrek Conference: Envisioning Future Media Environments*, 9–15.

Deuze, M. (2011). Media life. *Media, Culture, & Society, 33*, 137–148. doi: 10.1177/0163443710386518

Dorst, J. (1989). *The written suburb: An American site, an ethnographic dilemma*. Philadelphia: University of Pennsylvania Press.

———. (1999). *Looking west*. Philadelphia: University of Pennsylvania Press.

du Gay, P., Hall, S., Janes, L., Mackay, H., & Negus, K. (1997). *Doing cultural studies: The story of the Sony Walkman*. London: Sage Publications.

Elliott, S. (2010, June 15). TV ads coming to store aisles. *New York Times*. Retrieved from http://www.nytimes.com/

Ellul, J. (1964). *The technological society*. New York: Knopf.

———. (1980). *The technological system*. New York: Continuum.

———. (1990). *The technological bluff*. Grand Rapids, MI: William B. Eerdmans Publishing.

Evens, A. (2003). Concerning the digital. *Differences: A Journal of Feminist Cultural Studies, 14*, 49–77. doi:10.1215/10407391-14-2-49

Fernback, J. (2007). Beyond the diluted community concept: A symbolic interactionist perspective on online social relations. *New Media & Society, 9*, 49–68. doi: 10.1177/1461444807072417

Finneran, C., & Zhang, P. (2003). A Person-Artefact-Task (PAT) model of flow antecedents in computer-mediated environments. *International Journal of Human-Computer Studies, 59*, 475–496. doi: 10.1016/S1071-5819(03)00112-5

Fluss, D. (2015). Grow Motivation Through Gamification. *CRM Magazine, 19*, 4.

Frasca, G. (2003). Simulation versus narrative: Introduction to ludology. In M. Wolf & B. Perron (Eds.), *The video game theory reader* (pp. 221–236). New York: Routledge.

Friedberg, A. (1993). *Window shopping: Cinema and the postmodern*. Berkeley, CA: University of California Press.

Garfinkel, H. (1967). *Studies in ethnomethodology*. Englewood Cliffs, NJ: Prentice-Hall.

Geertz, C. (1983). *Local knowledge: Further essays in interpretive anthropology*. New York: Basic Books.

Gergen, K. (2000). *The saturated self*. New York: Basic Books.

———. (2002). The challenge of absent presence. In J. Katz & M. Aakhus (Eds.), *Perpetual contact: Mobile communication, private talk, and public performance* (pp. 227–242). Cambridge: Cambridge University Press.

Getting to know you; data. (2014, Sep 13). *The Economist, 412*, 5-S.7.

Giannetto, D., Chao, J., & Fontana, A. (2013). Gamification in a social learning environment. *Issues In Informing Science & Information Technology, 10*, 195–207.

Goggin, G. (2006). *Cell phone culture mobile technology in everyday life*. New York: Routledge.

Goggin, G. (2011). Ubiquitous apps: Politics of openness in global mobile cultures. *Digital Creativity, 22*, 148–159. doi: 10.1080/14626268.2011.603733

Gottschalk, S. (1995). Ethnographic fragments in postmodern spaces. *Journal of Contemporary Ethnography, 24*, 195–228. doi: 10.1177/089124195024002003

Graham, J. (2013, December 17). Apple reveals top downloads of 2013. *USA Today*. Retrieved from http://www.usatoday.com/

Green, G., & Kaufman, J. (2015). *Video games and creativity*. Elsevier: London.

Greenberg, J. (2015, June 8). Apple's news app takes aim at Facebook. *Wired*. Retrieved from wired.com

Grossberg, L. (1992). *We gotta get out of this place: Popular conservatism and postmodern culture*. New York: Routledge.

Gumpert, G., & Cathcart, R. (1986). *InterMedia: Interpersonal communication in a media world* (3rd ed). New York: Oxford University Press.

Habitica Gamifies. (2015 November 5). PR Newswire. *Reuters*. Retrieved from http://www.reuters.com/

Harms, J., & Dickens, D. (1996). Postmodern media studies: Analysis or symptom? *Critical Studies in Mass Communication, 13*, 210–227. doi: 10.1080/15295039609366976

Harnett, S. (2014, October 2). Why companies really want you to use an app for that. *Marketplace*. Retrieved from http://www.marketplace.org/

Harold, C. (2007). *OurSpace: Resisting the corporate control of culture*. Minneapolis, MN: University of Minnesota Press.

Hartmann, M. (2009). Everyday: Domestication of mediatization or mediatizatized domestication. In K. Lundby (Ed.), *Mediatization: concept, changes, consequences* (pp. 225–243). New York: Peter Lang.

Hartnett, K. (2015, June 24). 'Alice's Adventures in Wonderland' turns 150. *Boston Globe*. Retrieved from https://www.bostonglobe.com/

Heidegger, M. (1977). *The question concerning technology and other essays*. (W. Lovitt, Trans.). New York: Harper & Row Publishers (Original work published 1962).

Hestres, L. (2013). App neutrality: Apple's App Store and freedom of expression online. *International Journal of Communication, 7*, 1265–1280. doi: 1932–8036/20130005

Hjarvard, S. (2008a). The mediatization of religion: A theory of the media as agents of religious change. *Northern Lights, 6*, 9–26. doi: 10.1386/nl.6.1.9_1

———. (2008b). The mediatization of society. *Nordicom Review, 29*, 105–134.

Hymes, D. (1974). *Foundations in sociolinguistics: An ethnographic approach*. Philadelphia, PA: University of Pennsylvania Press.

Ikea. (2015). The new Ikea catalogue is here! Retrieved from http://www.ikea.com/ms/en_KR/catalogue-2015/index.html

Iñárritu, A.G. (Director). (2014). *Birdman, or (The Unexpected Virtue of Ignorance)*. Beverly Hills, CA: 20th Century Fox.

Ingram, M. (2015, September 21). TV engagement secrets from Twitter, everyone's favorite second screen. *Fortune*. Retrieved from http://fortune.com/

Introna, L. & Ilharco, F. (2004). The ontological screening of contemporary life: A phenomenological analysis of screens. *European Journal of Information Systems, 13*, 221–234. doi: 10.1057/palgrave.ejis.3000503

iOS. (2016). iOS 9: The most advanced mobile experience. Apple.com. Retrieved from http://www.apple.com/ios/

Jackson, J. (1990). "Deja Entendu": The liminal qualities of anthropological fieldnotes. *Journal of Contemporary Ethnography*, 19, 8–43. doi: 10.1177/089124190019001002

Jameson, F. (1991). *Postmodernism or the cultural logic of late capitalism*. Durham, NC: Duke University Press.

Jenks, C., & Neves, T. (2000). A walk on the wild side: Urban ethnography meets the flaneur. *Cultural Values, 4*, 1–17. doi: 10.1080/14797580009367183

Jin, S. A. (2011). "I feel present. Therefore, I experience flow:" A structural equation modeling approach to flow and presence in video games. *Journal of Broadcasting & Electronic Media, 55*, 114–136. doi: 10.1080/08838151.2011.546248

———. (2012). ''Toward integrative models of flow'': Effects of performance, skill, challenge, playfulness, and presence on flow in video games. *Journal of Broadcasting & Electronic Media, 56*, 169–186. doi: 10.1080/08838151.2012.678516

Jones, R. & Foust, C. (2008). Staging and enforcing consumerism in the city: The performance of othering on the 16th street mall. *Liminalities: A Journal of Performance Studies, 4*, Retrieved from http://liminalities.net/4-1/16thstreet.htm

Jones, K., Zagacki, K., & Lewis, T. (2007). Communication, liminality, andhope: The September 11th missing person posters. *Communication Studies, 58*, 101–121. doi: 10.1080/10510970601168780

Jonze, S. (Director). (2002). Adaptation [Motion Picture]. United States: Columbia.

Kellner, D. (2007). Media, spectacle, fear, and terrorism. *Media Development, 3*, 11–14.

Kien, G. (2009). *Global technography: Ethnography in the age of mobility*. New York: Peter Lang.

Knapp, K., Blair, L., and Mesch, R. (2014). *The gamification of learning and instruction fieldbook*. San Francisco: Wiley.

Korzybski, A. (1958). "Map is not the territory." *Science and sanity: An introduction to non-aristotelian systems and general semantics*. Brooklyn, NY: Institute of General semantics.

Koskinen, I. (2007). *Mobile multimedia in action*. New Brunswick, NJ: Transaction Publishers.

Krizek, R. (2007). Ethnography as postmodern practice: Accessing meaning in the ephemeral moments of events. Paper presented at the National Communication Association Annual Convention, Chicago, IL.

Krotz, F. (2009). Mediatization: A concept with which to grasp media and societal change. In K. Lundby (Ed.), *Mediatization: Concept, changes, consequences* (pp. 21–41). New York: Peter Lang.

Lacroix, C. & Westerfelhaus, R. (2005). From the closet to the loft: Liminal license and socio-sexual separation in *Queer Eye for the Straight Guy*. *Qualitative Research Reports in Communication, 6*, 11–19. doi: 10.1080/17459430500262109

Lakoff, G., & Johnson, M. (1980). *Metaphors we live by*. Chicago: University of Chicago Press.

Lanham, R. (2006). *The economy of attention: Style and substance in the age of information*. Chicago: University of Chicago Press.

Lash, S. (2002). *Critique of information*. Thousand Oaks, CA: Sage Publications.

LeBreton, G. (2009, July 20). Texas stadium's big screen tv makes it hard to watch the game. *Fort Worth Star-Telegram*. Retrieved from http://www.star-telegram.com

Lefebvre, H. (2003). Everyday life in the modern world. New Brunswick, NJ: Transaction Books.

———. (2002). *The critique of everyday life volume 2: Foundations for a sociology of the everyday*. New York: Verso.

Levitan, S. (Producer). (2009). Pilot [Episode of Television Series]. Modern Family. Hollywood, CA: 20th Century Fox Television.

Lindstrom, M. (2012). *Brandwashed*. New York: Crown Business.

Linklater, R. (Director). (2014). *Boyhood*. New York: IFC Productions.

Livingstone, S. (2009). On the mediation of everything. *Journal of Communication, 59*, 1–18. doi: 10.1111/j.1460-2466.2008.01401.x

Lorre, C. (2012). The beta test initiation [Television Series Episode]. *The Big Bang Theory*. Burbank, CA: Warner Brothers.

Lundby, K. (2009). Introduction: Mediatization as key. In K. Lundby (Ed.), *Mediatization: Concept, changes, consequences* (pp. 1–20). New York: Peter Lang.

Lynley, M. (2012, December 13). Meet Apple's top iPhone apps of 2012. *Wall Street Journal*. Retrieved from http://blogs.wsj.com/

Madden, M. (2014). Public perceptions of privacy and security in the post-Snowden era. Pew Research Center. Retrieved from http://www.pewinternet.org/2014/11/12/public-privacy-perceptions/

Madison, S. (2005). *Critical ethnography: Method, ethics, and performance*. Thousand Oaks, CA: Sage.

Magee, D. (2011, October 15). Steve Jobs day: Notable quotes from the late Apple co-founder, CEO. *International Business Times*. Retrieved from http://www.ibtimes.com/

Manjoo, F. (2014, January 14). Gamification of the office approaches. *Wall Street Journal*, B.1.

Manovich, L. (2001). *The language of new media*. Cambridge, MA: The MIT Press.

———. (2013). *Software takes command*. New York: Bloomsbury.

McLaren, P. (1997). *Revolutionary multiculturalism: Pedagogies of dissent for the new millennium*. Boulder, CO: Westview press.

Miller, G. (2014, September 11). Using smartphones to track our everyday moral judgments. *Wired Magazine*. Retrieved from http://www.wired.com/

Moscaritolo, A. (2015, September 2). These are the most popular iOS app, games ever. *PC Magazine*. Retrieved from http://www.pcmag.com/

Murphy, A. (2002). Organizational politics of place and space: The perpetual liminoid performance of commercial flight. *Text and Performance Quarterly, 22*, 297–316. doi: 10.1080/10462930208616175

Nakamura, J., & Csikszentmihalyi, M. (2001). The concept of flow. In C. R. Snyder & Shane J. Lopez (Eds.), *Handbook of positive psychology* (pp. 89–105). New York: Oxford University Press.

Nielsen. (2014, February 2). The U.S. Digital Consumer Report. The Nielsen Company. Retrieved from http://www.nielsen.com/us/en/insights/reports/2014/the-us-digital-consumer-report.html

———. (2015, January 20). Technology puts consumers in the driver's seat with connected cars. The Nielsen Company. Retrieved from http://www.nielsen.com/us/en/insights/news/2015/technology-puts-consumers-in-the-drivers-seat-with-connected-cars.html

No Ad. (2015). How it works. Retrieved from http://noad-app.com

Nolan, J., & McBride, M. (2014). Beyond gamification: Reconceptualizing game-based learning in early childhood environment. *Information, Communication & Society, 17*, 594–608. doi:10.1080/1369118X.2013.808365

Norman, D. (2013). *The design of everyday things*. New York: Basic Books.

Noy, C. (2009) On driving a car and being a family: An autoethnography. In P. Vannini (Ed.), *Material culture and technology in everyday life: Ethnographic approaches* (pp. 101–114). New York: Peter Lang.

NPR Staff. (2010, July 20). 'Hamlet's BlackBerry': To surf or not to surf? *National Public Radio*. Retrieved from http://www.npr.org/

Obrien, C. (2013). Louis CK hates cell phones. *Conan*. Retrieved from http://teamcoco.com/video/louis-ck-springsteen-cell-phone

Oremus, W. (2014, March 3). The smartphone wars are coming to your car. *Slate*. Retrieved from http://www.slate.com/

Ott, B. (2004). (Re)locating pleasure in media studies: Toward an erotics of reading. *Communication and Critical/Cultural Studies, 1*, 194–212. doi: 10.1080/14791420410001685386

Ott, B., & Lazic, G. (2013). The pedagogy and politics of art in postmodernity: Cognitive mapping and the bothersome man. *Quarterly Journal of Speech, 99*, 259–282. doi: 10.1080/00335630.2013.806816

Ott, B., & Mack, R. (2009). *Critical media studies: An introduction*. New York: Wiley.

Peterson, C. (2006). *A Primer in Positive Psychology*. New York: Oxford University Press.

Philipsen, G. (1992). *Speaking culturally: Explorations in social communication*. Albany, NY: State University of New York Press.

Pierce, D. (2016, June 12). How Apple taught the world to smartphone. *Wired*. Retrieved from wired.com

Pieterse, J. (2001). Hybridity, so what?: The anti-hybridity backlash and the riddles of recognition. *Theory, Culture, & Society, 18*, 219–245. doi: 10.1177/026327640101800211

Pink, S. (2012). *Situating everyday life*. London: Sage.

Postman, N. (1986). *Amusing ourselves to death: Public discourse in the age of show business*. New York: Penguin Books.

———. (2000). The humanism of media ecology. *MEA Proceedings*. Retrieved from http://mediaecology.org/publications/MEA_proceedings/v1/humanism_of_media_ecology.html

Raphael, C., Bachen, C., & Hernández-Ramos, P. (2012). Flow and cooperative learning in civic game play. *New Media Society, 14*, 1321–1338. doi: 10.1177/1461444812448744

Reisinger, D. (2016, January 11). Apple news has one major glitch. *Fortune*. Retrieved from http://fortune.com/

Re+Public. (March 2, 2015). Re+Public: Reimagining Public Space. Retrieved from http://republiclab.com/

Reese, H. (2014, July 14). Digital sanctuaries. Retrieved from http://cityofasylum.org/digital-sanctuaries-a-jazz-poetry-sidewalk/

Rheingold, H. (1993). *The virtual community: Homesteading on the electronic frontier*. Reading, Mass: Addison-Wesley Pub. Co.

Rodriguez, S. (2013, August 20). Why new Waze feature will keep Google maps miles ahead of rivals. *Los Angeles Times*. Retrieved from http://articles.latimes.com/

Sandywell, B. (2004). The myth of everyday life: Toward a heterology of the ordinary. *Cultural Studies, 18*, 160–180. doi: 10.1080/0950238042000201464

Scannell, P. (2000). For-anyone-as-someone structures. *Media, Culture & Society, 22*, 5–24. doi: 10.1177/016344300022001001

Scott, S. (2009). *Making sense of everyday life*. Cambridge: Polity Press.

Schulz, W. (2004). Reconstructing mediatization as an analytical concept. *European Journal of Communication, 19*, 87–101. doi: 10.1177/0267323104040696

Schwartz, T. (2015, November 29). Addicted to distraction. *New York Times.* Retrieved from http://www.nytimes.com/

Seiler, J. (2015). Re + Public App. Retrieved from https://itunes.apple.com/us/app/re+public/id683353374?mt=8

Seinfeld, J. (2014). The Tonight Show with Jimmy Fallon. Retrieved from https://www.youtube.com/watch?v=xR1ckgXN8G0

Sheller, M., & Urry, J. (2006). *Mobile technologies of the city.* New York, NY: Routledge.

Shernoff, D. J., & Csikszentmihalyi, M. (2009). Flow in schools: Cultivating engaged learners and optimal learning environments. In R. Gilman, E. S. Huebner, & M. Furlong (Eds.), *Handbook of positive psychology in schools* (pp. 131–145). New York: Routledge.

Sherry, J. L. (2004). Flow and media enjoyment. *Communication Theory, 14*, 328–347.

Shields, R. (2000) Performing virtualities: Liminality on and off the 'Net'. Retrieved from http://virtualsociety.sbs.ox.ac.uk/events/pvshields.htm

SI Wire (2016, January 15). Inglewood stadium to have 120 yard video board. Sports Illustrated. Retrieved from http://www.si.com/

Silverstone, R. (2002). Complicity and collusion in the mediation of everyday life. *New Literary History*, 33, 761–780.

Simonite, T. (2014, February 12). Sell your personal data for $8 a month. *MIT Technology Review*. Retrieved from https://www.technologyreview.com/

Siska, W. (1979). Metacinema: A modern necessity. *Literature Film Quarterly, 7*, 285–290.

Slocombe, W. (2013). *Nihilism and the sublime postmodern*. Routledge: New York.

Smith, A. (2010, October 14). Pew Research Center. Americans and their gadgets. Retrieved from http://www.pewinternet.org/Reports/2010/Gadgets.aspx

———. (2015, April 1). U.S. smartphone use. Pew Research Center. Retrieved from http://www.pewinternet.org/2015/04/01/us-smartphone-use-in-2015/

Soukup, C. (2013). The postmodern ethnographic flaneur and the study of hyper-mediated everyday life. *Journal of Contemporary Ethnography, 42*, 226–254. doi: 10.1177/0891241612461278

St. John, G. (2004). Introduction. In G. St. John (Ed.), *Rave culture and religion* (pp. 1–15). London: Routledge.

Starr, M. (2009, July 22). More TVs than humans. *New York Post*. Retrieved from http://www.nypost.com/

Stearns, P. (2003). *Anxious Parents: A history of modern childrearing in America.* New York: New York University Press.

Stelter, B. (2009, March 26). 8 hours a day spent on screens, study finds. *New York Times*. Retrieved from http://www.nytimes.com/

Stenros, M., & Waern, A. (2009). Pervasive game design strategies. In M. Montola, J, Stenros, and A. Waern (Eds.), *Pervasive games: Theory and design* (pp. 137–157). Amsterdam: Morgan Kaufamann Publishers.

Sussman, H. (2003). Between the registers: The allegory of space in Walter Benjamin's arcades project. *Boundary, 2*, 169–190.

Taylor, M. (2012). *Musical theatre, realism and entertainment*. Ashgate Publishing: Burlington, VT.

Taylor, C. (2001). Intellectual capital. *Computerworld, 35*, p. 51.

Teece, D. (2000). *Managing intellectual capital: Organizational, strategic, and policy dimensions*. New York: Oxford University Press.

Thomas, A. (2011). *The business wisdom of Steve Jobs: 250 quotes from the innovator who changed the world*. New York: Skyhorse Publishing.

Timberg, S. (2015, November 30). The internet is scrambling our brains. *Salon*. Retrieved from http://www.salon.com/

Turner, V. (1975). Symbolic studies. *Annual Review of Anthropology, 4*, 145–161. Retrieved from EBSCO*host*.

———. (1979). *Process, performance, and pilgrimage: A study in comparative symbology*. New Delhi: Concept Publishing.

———. (1980). Social dramas and stories about them. *Critical Inquiry, 7*, 141–168.

———. (1982). *From ritual to theatre: The human seriousness of play*. New York: Performing Arts Journal Publications.

Tryon, C. (2009). *Reinventing cinema: Movies in the age of media convergence*. New Brunswick, NJ: Rutgers University Press.

US Census. (2014). Computer and Internet trends in America. United States Census Bureau. February 4, 2014.

US Department of Transportation. (2016). Facts and Statistics: Distracted driving. Retrieved from http://www.distraction.gov/stats-research-laws/facts-and-statistics.html

Utterback, J. (1996). *Mastering the dynamics of innovation*. New York, NY: Harvard-Business Review Press.

Valentino-Devries, J. (2011, August 24). Steve Jobs best quotes. *Digits: Tech News and Analysis from The Wall Street Journal*. Retrieved from http://blogs.wsj.com/digits/

Van Maanen, J. (1988). *Tales of the field: On writing ethnography*. Chicago: University of Chicago Press.

Vannini, P. (2009). *Material culture and technology in everyday life: Ethnographic approaches*. New York: Peter Lang.

Vannini, P., Waskul, D., Gottschalk, S., and Rambo, C. (2010). Sound acts: Elocution, somatic work, and the performance of sonic alignment. *Journal of Contemporary Ethnography, 39*, 328–353. doi: 10.1177/0891241610366259

Virilio, P. (1991). *Lost dimension*. Brooklyn, NY: Semiotext.

———. (2002). *Crespuscular dawn*. Brooklyn, NY: Semiotext.

———. (2004). *The Paul Virilio reader*. New York: Columbia University Press.

Visser, M. (2007, August 3). Sprint commercial: Dreams [Video File]. Retrieved from https://www.youtube.com/watch?v=EvELLlfcBKQ

Wakabayashi, D. (2016, January 6). Apple's App Store sales hit $20 billion. *Wall Street Journal*. Retrieved from http://www.wsj.com/

Wallace, K. (2015, November 3). Teens spend a 'mind boggling' 9 hours a day using media, study says. *CNN*. Retrieved from http://www.cnn.com/

Walsh, V. (2003). *Dora the Explorer* and the Lost City [Television Series Episode]. *Dora the Explorer*. Hollywood, CA. Paramount.

Weber, R., Tamborini, R., Westcott-Baker, A., and Kantor, B. (2009). Theorizing flow and media enjoyment as cognitive synchronization of attentional and reward networks. *Communication Theory, 19*, 397–422. doi: 10.1111/j.1468-2885.2009.01352.x

Weibel, D., Wissmath, B., Habegger, S., Steiner, Y., & Groner, R. (2007). Playing online games against computer vs. human-controlled opponents: Effects on presence, flow, and enjoyment, *Computers in Human Behavior*. Retrieved from http://www.sciencedirect.com/

White, G. (2015, February 25). Why the gap between worker pay and productivity is so problematic. *The Atlantic*. Retrieved from http://www.theatlantic.com/

Wingfield, (2012, August 25). Jury awards $1 billion to Apple in Samsung patent case. *New York Times*. http://www.nytimes.com/

Wood, A. (2005). ''What happens [in Vegas]'': Performing the post-tourist flaneur in''New York'' and ''Paris.'' *Text and Performance Quarterly, 25*, 315–333. doi: 10.1080/10462930500362403

———. (2009). *City ubiquitous: Place, communication, and the rise of omnitopia.* New York: Hampton Press.

Worland, J. (2014, November 6). Why people text and drive even when they know it's danger. *Time Magazine*. Retrieved from http://time.com/

Index

About the Author

Charles Soukup is a professor of communication at the University of Northern Colorado. His research and teaching emphasize the role of technology and media in the formation of culture and community. On a more personal note, Charles lives in Denver with his wife (Christina), daughter (Rose), and son (Henry).

www.ingramcontent.com/pod-product-compliance
Ingram Content Group UK Ltd.
Pitfield, Milton Keynes, MK11 3LW, UK
UKHW041626031026
15587UKWH00001B/83

9 781498 539609